TRAVELS WITH TRILOBITES

TRAVELS WITH TRILOBITES

ADVENTURES IN THE PALEOZOIC

ANDY SECHER

Forewords by Niles Eldredge, Mark Norell, and Kirk Johnson

Columbia University Press   *New York*

Columbia University Press
Publishers Since 1893
New York   Chichester, West Sussex
cup.columbia.edu

Library of Congress Cataloging-in-Publication Data
Names: Secher, Andy, author.
Title: Travels with trilobites : adventures in the paleozoic / Andy Secher.
Description: New York : Columbia University Press, [2022] | Includes index.
Identifiers: LCCN 2021037496 (print) | LCCN 2021037497 (ebook) | ISBN 9780231200967 (hardback) |
    ISBN 9780231553865 (ebook)
Subjects: LCSH: Trilobites. | Extinction (Biology)
Classification: LCC QE821 .S43 2022 (print) | LCC QE821 (ebook) | DDC 565/.39—dc23
LC record available at https://lccn.loc.gov/2021037496
LC ebook record available at https://lccn.loc.gov/2021037497

Cover design: Milenda Nan Ok Lee

COVER IMAGE: *PARACERAURUS ACULEATUS* (EICHWALD, 1857)

Upper Ordovician, Caradocian; Kukruse Regional Stage; Viivikonna Formation, Alekseevka Quarry; St. Petersburg region, Russia; 10.4 cm

TITLE PAGE IMAGE: *HOMOTELUS BROMIDENSIS* (ESKER, 1964)

Ordovician; Criner Hills; Carter County, Oklahoma, United States; largest trilobite: 6.2 cm

At this site, thousands of complete *Homotelus* specimens have been found in tightly packed alignment.

CONTENTS PAGE IMAGE: *NEVADIA PARVOCONICA* FRITZ, 1992

Lower Cambrian; Poleta Formation; Esmeralda County, Nevada, United States; largest trilobite: 3 cm

Examples of this unusual species were uncovered in the early years of the twenty-first century. So far, no more than half-a-dozen articulated specimens have been found.

PAGE XI IMAGE: *EOBRONTEUS LUNATUS* (BILLINGS, 1857)

Middle Ordovician; Trenton Group; Ottawa, Ontario, Canada; 7.2 cm

Scutellid trilobites are exceedingly unusual in the Trenton Group, especially when they appear as complete specimens.

This book is dedicated to everyone, everywhere, who has ever taken a moment to contemplate the wonders of trilobites. It doesn't matter if you're a casual collector who perhaps possesses a lone *Elrathia kingii* perched atop an office shelf or an advanced enthusiast whose free time (and spare cash) is happily spent in the pursuit and procurement of these often-elusive Paleozoic prizes. Or perhaps you're just someone interested in learning more about one of the most captivating, diverse, and strangely beautiful organisms ever to inhabit our planet's primal seas—the trilobite.

I would also like to mention the support, loyalty, and guidance supplied by Jodi Summers. Throughout the years she has remained an eager and welcomed participant in many of my trilobite-related escapades to all corners of the globe. Jodi's love of travel and unmatched spirit of adventure have made her an invaluable and much appreciated companion, whether it be in the heat of the Arizona desert or the late spring chill of a French mountainside.

Contents

2. ORDOVICIAN PERIOD: 485–444 MILLION YEARS AGO

3. SILURIAN PERIOD: 444–419 MILLION YEARS AGO

4. DEVONIAN PERIOD: 419–359 MILLION YEARS AGO

5. CARBONIFEROUS/PERMIAN PERIODS: 359–252 MILLION YEARS AGO

6. TRILOBITE THOUGHTS AND OBSERVATIONS

On the Space-Time Road with Trilobites

G. ARTHUR COOPER, a famous paleontologist of the mid-twentieth century, opened one of his papers on the sequence of Middle Devonian fossiliferous rocks in New York State with a paean of praise for the internal combustion engine. Less than 100 years later, that particular clever invention in large measure fuels the climate change that threatens the very existence of life on Earth. But back then it was the pure joy and exhilaration of taking your car and going pretty much wherever you wanted, whenever you wanted, to visit often remote rock outcrops and explore them for their amazing fossilized remains. No more trains, no more buggy rides—and no more long hikes across open fields and along dirt trails to get to those lakeshore, stream bank, and cliffside outcrops that nature had to offer. Efficiency replaced what we romantically are tempted to think of as the more bucolic, relaxed travel days of yore.

The pace of discovery duly quickened. And the responsible, determined acquisition of fossils—for scientific analysis as well as the equally profound emotional and aesthetic thrill of simply seeing an exquisite form of ancient life—has mushroomed logarithmically. Millions of fossils are now safely housed in museum collections or on display for all who wish to share in the mystery, thrill, and knowledge that fossils have to offer. Fossils are of the Earth, faithful vestiges of what life has been like in its long evolutionary journey over the past three-billion-plus years.

Increasingly, the private sector has played a determined, informed role in the ongoing exploration, collection, and meticulous cleaning ("preparation") and proper storage of fossils. The assembly, care, and feeding of collections of rare, beautiful, and scientifically important specimens (I believe all specimens, no matter how beat up and ugly, are scientifically important!) is sometimes done better in dedicated private hands than in museums, with their limited resources and more diverse programmatic charges.

Best, of course, is when private collectors are affiliated with museums that aid in their fossils' long-term survival, eventual scientific study, and in their public display, thus spreading their aesthetic, emotional, and scientific significance. Andy Secher's book is a rare example of the gift of beauty and knowledge of a passionate private collector who has gone to considerable effort to explain and display his dearly loved and carefully curated specimens to the world at large. His lucid text and fabulous photos take us along the space-time road of trilobite discovery. It is a succinct and fascinating retracing of the actual space-time road that trilobites themselves traveled in their evolutionary/ecological journey—a journey that started at what we now define as the base of the Cambrian Period. Trilobites are the very icon of the famous Cambrian Explosion, that breathtakingly quick (only 10 million years!) burst of evolutionary diversification of complex forms of animal life still with us today—albeit in altered form.

Extinction is a rule of life. Without it, life would not have evolved still further. Trilobites are now long gone although they survived global extinction events at the end of the Cambrian, then the Ordovician, and still later, the Upper Devonian. Already dwindling and on the ropes in the boxing ring of life, trilobites finally succumbed in that greatest mass extinction (so far!) of all at the end of the Permian Period. Trilobites frame the entire Paleozoic Era, from beginning to end. All the (nearly) 300 million years of it.

So why, we ask, do trilobites often form the focal point for a collection, for a book, for an exhibition, or for that matter for a scientific career? Why not, say, the more romantically famous dinosaurs? Sure, collecting dinosaurs presents logistical and financial difficulties that transcend collecting cars or even tubas. But still . . . I have long thought that trilobites attract their devoted following because they actually *look like an animal*. Trilobites have a head (cephalon), usually with well-developed eyes. They have a segmented middle part (the thorax) that is flexible and allows them to roll up for defense, just as modern-day crustaceans such as garden pill bugs can and do. And their fused-segment tails (pygidium) bring up the rear. In other words, you can relate to a trilobite as a fellow animal far more easily than to the usually scrappily incomplete remains of even the most compelling of the dinosaurs.

It is their eyes, I think, that make the real difference. It is obvious, with a simple glance of our own eyes, that trilobites could *see*. They could see while they traveled along their nearly 300-million-year-long space-time road. And now, thanks to this splendid book, we can better see them taking that road. And we can marvel at it all.

Niles Eldredge
Curator Emeritus
Division of Paleontology
American Museum of Natural History

FOSSILS ARE OUR PLANET'S MEMORY, and they stick hard in our memories. Those people who are lucky enough to have found fossils often remember their first fossil for life. Leaving dinosaurs aside for the moment, two types of fossils seem to appeal more than others: ammonites and trilobites. Both are extinct, diverse, and beautiful; both are poster children for extinction; and both have fans who devote their lives to collecting them. Andy Secher has one of the most extreme cases of trilobite-o-philia ever documented. I met Andy about 30 years ago when his disease was already well-advanced. As a paleontologist who knows from personal experience what it means to love fossils, I recognized Andy as a kindred spirit and was delighted to get to know him.

Over the last three decades, I have watched the world of trilobites expand as new sites are discovered, classic sites are rediscovered, more aggressive forms of excavation are deployed, and most important, the craft of fossil preparation is pushed to new levels. The widespread use of air-scribes, needles, and intense patience has revealed the world of three-dimensional trilobites, and there seems to be no end to the possibilities as spines, eyes, and other exquisite body parts are revealed after hundreds of millions of years of entombment.

These efforts were not led by museums but by individuals who fed a rapidly growing global collector's market for trilobites. A trend that started as a hobbyist's obsession for the perfect *Elrathia*, *Phacops*, or *Flexicalymene* has exploded with the rise of fossil shows and with the exploitation of fossil sites in Morocco, Russia, the United States, and

China. People now routinely pay more than $10,000 for a high-quality "bug." The Anthropocene has encountered the Paleozoic, and Andy has been there for the whole transition.

As a result, the vast majority of the world's best trilobites collected in the last 30 years and are in private hands, not in museums. Museum collections hold the results of the geological surveys of the Paleozoic over the last two centuries, and most of their trilobites are represented by partial specimens, the results of more than two million centuries of molting. Now, specimen by specimen or collection by collection, these new fancy trilobites are slowly making their way into museum collections and exhibitions.

Fortunately for science and for trilobite lovers, Andy has forged a partnership with the American Museum of Natural History to create a trilobite website—which has led to this fantastically comprehensive book.

Kirk Johnson
Sant Director
National Museum of Natural History
Smithsonian Institution
Washington, D.C.

The Collector's Gene

AS A PALEONTOLOGIST, I understand trilobites for their scientific importance. They are one of the most recognizable forms of prehistoric life (only overshadowed by dinosaurs). But as a collector (not of trilobites!), I can also understand them from another perspective; one that is purely aesthetic.

Some people have said that there is a collectors' gene. Part of my life as a professional paleontologist involves developing collections from excavation, gift, or purchase. I have absolutely no interest in owning fossils, and with few exceptions don't really understand why you would want them in your house. But I understand collectors. One of my deceased colleagues made perhaps the world's greatest collection of airplane vomit bags. Without giving too much away, it deals with things that hang on walls and are pretty nice to look at.

In the collecting world, there are three classes: synoptic collectors, accumulators, and taste makers who only want the best of the best. Andy Secher is an amalgam of all three. His trilobite collection encompasses most of the known species. Needless to say, he has a lot of trilobites—one of the largest collections in the world and premier examples of many species. Of course, as a nonspecialist I would not know this, but I have seen Andy pretty excited over what looks like a pencil smudge on a piece of slate.

A visit to Andy's home shows an unusual fusion of rock and roll paraphernalia (tour guitars and jackets, and a lot of albums). The shelves of albums, sagging under their weight, are more depleted every visit. Vinyl is going the way of used toys to make room for more "bugs." I wonder if they feel somewhat like the toys in the final *Toy Story*—sad, neglected, and given away. Probably a happy ending though, because vinyl is coming back and there is undoubtedly some 14-year-old metal head listening to them now.

Finally, when we think about fossils, usually the image of a khaki-clad nerd in a museum comes to mind. One look at Andy and his collection dispels all of that. Instead of clinging to every specimen, his generosity has been prodigious in providing the AMNH with important scientific and display specimens. This book is a testament to that. Not in the sense of the science but as a great look into the world of a cadre of specialized collectors. It provides insight into all who love objects.

MARK NORELL
MACAULAY CURATOR OF PALEONTOLOGY
CHAIRMAN DIVISION OF PALEONTOLOGY
AMERICAN MUSEUM OF NATURAL HISTORY

GABRIELLUS KIERORUM (CORBACHO AND LOPEZ-SORIANO, 2013)

Lower Cambrian; Rosella Formation, Atan Group; Dease River, Northern British Columbia, Canada; 11.2 cm

This is a particularly attractive example an unusual species exhibiting significant predation on the right side of its thorax.

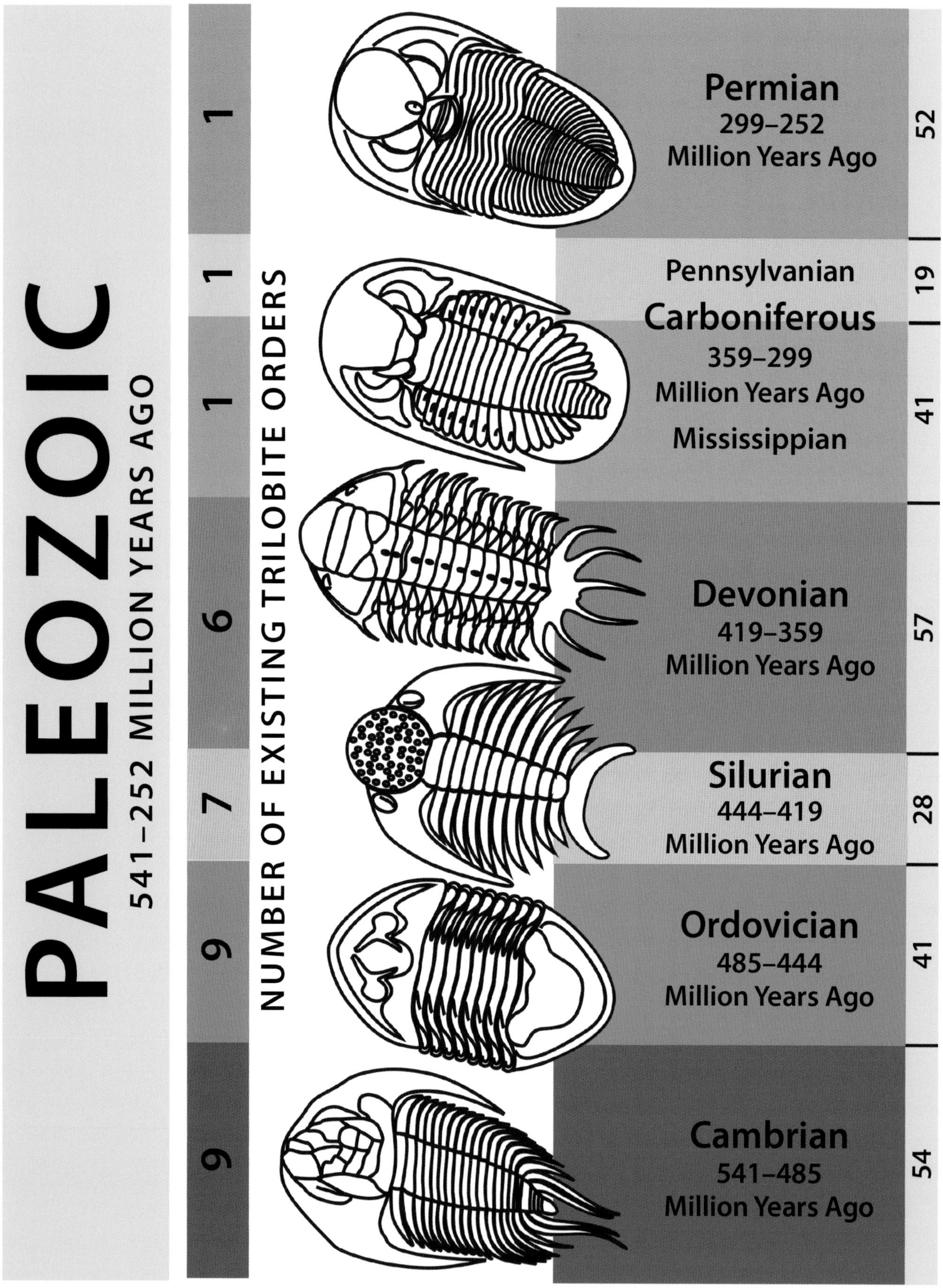

PALEOZOIC

541–252 MILLION YEARS AGO

NUMBER OF EXISTING TRILOBITE ORDERS

1

1

1

6

7

9

9

Permian
299–252
Million Years Ago

52

Pennsylvanian
Carboniferous
359–299
Million Years Ago
Mississippian

19

41

Devonian
419–359
Million Years Ago

57

Silurian
444–419
Million Years Ago

28

Ordovician
485–444
Million Years Ago

41

Cambrian
541–485
Million Years Ago

54

MILLIONS OF YEARS

Time chart designed by Mark Ault

PREFACE
Why Trilobites?

WHETHER YOU REVEL in their timeless Paleozoic permutations or merely view them as bizarre remnants of an incredibly distant age, it's difficult to deny that trilobites are fascinating creatures. These fossilized invertebrates hold the distinction of being among our planet's first forms of complex animal life, arthropods that initially appeared not long after the Earth had emerged from a semifrozen state and the oceans were ripe for the blooming of biodiversity.

By the time the Cambrian Period began producing an incredible array of unusual and previously unseen fauna some 521 million years ago, trilobites were already advanced organisms possessing hard protective exoskeletons, complex digestive systems, and highly developed eyes. Although their origins remain cloaked in evolutionary mystery, all signs indicate that the rootstock of the trilobite line stretches into an even earlier time in the planet's primordial past. Yet from the moment their calcified carapaces began filling the fossil record soon after the dawn of the famed Cambrian Explosion, trilobites were the omnipresent monarchs of the world's ancient seas.

These strangely beautiful life-forms would eventually evolve into more than 25,000 scientifically recognized species. And prior to their demise at the end of the Permian Period, they would rule the waves—or at least the shoals, shelves, and reefs lurking under those waves—for the next quarter-billion years.

ALLOLICHAS HALLI (FOERSTE, 1888)
Ordovician; Cincinnatian Series; Mt. Orab, Ohio, United States;
2.5 cm

This classic midwestern lichid represents one of the most sought-after U.S. trilobites. Disarticulated pieces are relatively common finds at this site, but complete specimens rank among the true treasures of the North American fossil landscape.

Around the globe, paleontologists have named entire geologic horizons for the prolific, frequently disarticulated exoskeletons of these long-gone ocean inhabitants. Indeed, it would be hard to find a place on Earth where the distinctive debris of trilobites is not preserved in various slates, shales, or sandstones. Trilobites evidently grew quite rapidly and, much like modern arthropods, molted numerous times each year. Thus, an overwhelming percentage of their fossilized remains are not of the deceased animals themselves but of their cast aside, and often fragmentary, external armor.

The fact is that trilobites have been uncovered just about everywhere that sedimentary outcrops of the right age exist, from Morocco's imposing Atlas Mountains, to Bolivia's soaring Altiplano, to the serrated escarpments surrounding California's Death Valley. They have been found in the desolate Siberian steppes, on the shores of frigid Scandinavian islands, in the bustling suburbs of major North American cities—and virtually every other Paleozoic place in between!

From present-day England to Canada, from Australia to China, from Greenland to the Czech Republic, the fossil record provides ample evidence that trilobites—or "bugs," as they are fondly called by those who find themselves inexorably drawn to these ancient arthropods—filled virtually every saltwater habitat of their primeval world.

In the Paleozoic, the roughly 290-million-year era that ran from the beginning of the Cambrian through the end of the Permian, the surface of our planet looked radically different than it does today. The continents we now instantly recognize by their distinctive shapes and familiar global alignment had not yet been transported to their current lithospheric locations via the geological phenomenon of plate tectonics. Instead, they were often packed together in nondescript clusters in the Earth's Southern Hemisphere, with one or at times two major landmasses dominating what was otherwise a water world.

Significant sections of what we presently view as the continents of North America, South America, Asia, Afric, Australia, Europe, and even

PLACOPARINA SEDGWICKI (MCCOY, 1849); SHELVENSIS (HUGHES, 1969)
Middle Ordovician, Lower Llanvirn Series, Abereiddian Stage
Shelve Formation, Hope Shale Member, Artus Biozone,
Whitsburn, Shropshire, England, UK; 5.1 cm

Beautiful and crisp preservation distinguishes this remarkable example of a relatively common species from the famed English Midlands.

Antarctica were then below sea level, allowing marine forms like trilobites to become fossilized in locations now hundreds of kilometers from the nearest prominent body of water.

Clearly, by any measurable means, trilobites were among the most important and abundant inhabitants of early Earth. In all honesty, however, despite their worldwide distribution and the essential, yet often overlooked, role they played in the development of life on our planet—possibly being the initial creatures to venture, albeit briefly, out of the seas—it is difficult for even the most ardent trilobite enthusiast to explain the almost mystical allure that these amazing organisms hold over both those who study and collect them.

Perhaps it is the fact that the fossils of so many unknown trilobite species are still lurking out there somewhere, just waiting for a determined explorer to free them from their eons-old rock encasements. Perhaps it is the incredible antiquity of these primal beasts, or the notion that they represent one of Earth's first successful experiments with complex organic matter. Or perhaps it is simply that in the opinion of a small but significant segment of the world's population, trilobites are just plain "cool." Featuring an astonishing assemblage of shapes, colors, and sizes—ranging from diminutive Cambrian ptychopariids that rarely exceeded a millimeter in length to giant species of Ordovician asaphids that attained proportions of more than 70 centimeters—it can be asserted that no other dwellers of our planet's deep-blue seas have ever displayed the diversity and the longevity of these singularly distinctive examples of Paleozoic life.

Trilobites have existed—either as living animals or fossilized forms—for more than half a billion years, but the collecting of these extraordinary organisms is a relatively recent occurrence. Many classic geology books reference certain erudite Europeans being aware of "frozen locusts," as they were then called, as far back as the eighteenth century. Other tomes explain that southwestern Native American tribes have long treated the *Elrathia kingii* carapaces they encountered with the reverence of religious artifacts, and some volumes may mention that as early as the tenth century noble houses in China proudly presented trilobite fossils as pieces of natural art.

It is also believed that Egyptian royalty flaunted trilobite fossils as symbols of power and prestige nearly 4,000 years ago. In fact, well-worn and apparently highly prized trilobite specimens have been found among the personal possessions of ice age humans discovered in European cave sites dating back more than 15,000 years. Even President Thomas Jefferson exhibited a small trilobite as part of his celebrated "cabinet of curiosities" scientific collection.

Other than these select and often fragmentary morsels of fossiliferous information—unquestionably highlighted by the pioneering nineteenth-century research and subsequent writings of the American paleontologist Charles Walcott, the British geologist Sir Roderick Murchison, and the French naturalist Joachim Barrande—historical references to trilobites remain rather obscure. We know, for example, that sociable Englishmen could frequently be found foraging for trilobites almost two centuries ago while they strolled through the Silurian deposits near the town of Dudley. And at a similar time, quarry workers in the fossil-rich sediments of what was then Bohemia were noted for their willingness to find, piece together, and then sell what were basically disassociated trilobite parts, in the process creating chimera-like monstrosities unknown to either science or the Cambrian seas.

Yet despite such past, fleeting moments of paleontological notoriety and renown, it seems that the focused collecting of these absorbing invertebrates—especially by amateur enthusiasts—is a distinct by-product of late-twentieth-century technology. It's been during the last 50-plus years that the advent of nimble off-road vehicles, powerful rock-moving machinery, and sensitive GPS trackers have made the exploration and excavation of new trilobite-bearing locations more accessible to both adventurers and academics. In addition, the emergence of small, affordable, air abrasive units has made detailed cleaning of any subsequent finds increasingly convenient and the resulting preparation more spectacular.

The most pronounced consequence of it all as we continue hurtling through the twenty-first century is that trilobites have begun to exert an ever-more forceful grip upon the hearts, minds, and pocketbooks of thousands of Paleozoically inclined people around the globe. Websites dedicated to either the scrutiny or sale of these engaging creatures have emerged as invaluable resources for both scientists and collectors. Indeed, one notable Facebook page, which provides a continually updated "all-things-trilobite" forum, recently bragged about being comprised of an active community of over 10,000 members—perhaps not quite up to Kardashian-clan standards, but still rather impressive considering its less than mainstream subject matter.

For many fossil enthusiasts, these rock-encased remnants now rank as the most significant of primordial faunal forms, surpassing even the hallowed dinosaur in their paleontological appeal. In recent years, feature stories focused on the collectability of trilobites have appeared everywhere, from the *New York Times* to *Forbes* magazine, and at the same time these astounding arthropods have continued to make ubiquitous appearances across the internet as well as in major natural history auctions held from London to Los Angeles. In loose conjunction with these developments, exciting trilobite-centric museum displays have recently opened in New York, Prague, Washington, D.C., Cancun, Barcelona, Houston, and Casablanca,

with each further spotlighting the surprisingly global allure exhibited by these 500-million-year-old marine relics.

Thanks to this intriguing combination of factors—including their easy availability for both study and sale, their often outlandish appearance, the amazing tales told by their fossilized remains, and the incredible duration of their passage through Earth history—trilobites, in all their multisegmented glory, represent one of early life's most captivating efforts. The quest here is to further reveal why, although if you've gotten this far, you may already know the answer!

In *Travels with Trilobites*, I invite you to come along on fossil-seeking sojourns to more than a score of paleontological strongholds found across the planet. Together we'll venture to Alnif, Morocco, the trilobite hub perched on the edge of the Sahara Desert . . . La Paz, located sky-high in the Bolivian Altiplano . . . the Sakha Republic, deep in the heart of the Siberian wilderness . . . and Kangaroo Island, the colorful destination situated off the coast of South Australia.

We'll also visit a multitude of somewhat more accessible trilobite-rich repositories, such as the renowned Rochester Shale quarry in upstate New York, the bountiful Haragan Formation deposits of Oklahoma, the historic Wenlock outcrops of England, the legendary Burgess Shale layers of British Columbia, and the rugged cliff-side trilobite beds of Anticosti Island, situated smack in the middle of Quebec's busy St. Lawrence Seaway.

And if that's not enough to satiate your trilobite-targeted cravings, we'll go on an eye-opening trip behind the scenes at one of North America's leading natural history museums and stop off at the famed fossil show that takes place each winter in Tucson, Arizona. There we will come face to cephalon with a choice selection of spectacular trilobite specimens, each designed to test the bounds of both your knowledge and your bank account.

In considered contrast to the profusion of scientifically inclined textbooks, dissertations, tomes, and treatises that have addressed the subject of the "World's Favorite Fossil Arthropod," the objective of *Travels with Trilobites* is to blend fact, fun, and fossiliferous content into a package of pure Paleozoic infotainment.

From the inception of this book, my avowed goal has been to appeal directly to the dedicated collector and the curious layperson—although my sincere hope is that there is also just enough hardcore scientific "stuff" presented to satisfy the oft-finicky tastes of any accredited academic. Through a series of globe and time-spanning essays, my intent is to go where no trilobite book has gone before, providing an interesting, enlightening, and at times even amusing look at trilobite hunters, collectors, preparators, and researchers, many of whom you'll hear from in their own words.

Interspersed among the various "location" chapters that serve as this work's cornerstone are more than two dozen concise reports focusing on such diverse yet related topics as trilobite eyes, the Paradoxides Paradox, ventral preservation, trilobite descendants (spoiler alert, there aren't any), trilobites in history, and the strangest trilobites (the latter of which affords many candidates). In chapter 6 you will encounter a series of thoughts and observations, each written to furnish a revealing perspective on everything from trilobite values and specimen preparation to curating a collection, international fossil rules and regulations, and even fake trilobites.

MACROPYGE SP. CF. CHERMII STUBBLEFIELD, 1927

Lower Ordovician; Tremadoc Series, Cressagian Stage; Mawddach Group, Dol-Cyn-Afon Formation; Nant-y-Ceunant, near Dolgellau; Gwynedd County, North Wales, UK; 6.2 cm

This trilobite represents one of the strangest genera in the trilobite class. Note the slight torque of the specimen on its slate matrix.

Of course, plenty of space is allotted to the trilobites themselves. There are hundreds of high-resolution, full-page color images featuring some of the most incredible specimens ever found—many exceptional examples of relatively familiar genera as well as one-of-a-kind trilobite treasures.

Please understand that the primary mission here has never been to create an all-encompassing work presenting photos and descriptions of every trilobite ever discovered, or even of every species seen within each of the featured locations. My goal has been to create a proper photographic outlet through which to both enhance this book's editorial content and showcase some of the most noteworthy trilobites ever unearthed. The specimens that have been selected—chosen from literally thousands of candidates to illustrate their elegantly eccentric configuration and often fanciful appearance—serve to promote the morphological variance, scientific significance, and collectible appeal that these intriguing invertebrates so effortlessly exude.

Feel free to approach the contents of this book in any manner you prefer. Each chapter can be perused in the order presented, or you can proceed by picking and choosing the subjects, concepts, or locations you find particularly engaging. Indeed, each editorial section of *Travels with Trilobites* has been designed to stand alone and apart from any surrounding text even as it simultaneously adds a key component to the timeless trilobite tale being told.

You may even decide initially to skip the arthropod-obsessed story line and focus solely on the eye candy display of trilobite photos that appear on virtually every page. Whether you choose to intently study the morphological details of each photographed specimen, mentally digest every carefully considered word, or just aspire to embrace trilobites as bizarrely beautiful creatures, I hope this book will remove any lingering doubt that these astonishing organisms—the planet's original "rock

**CERAURUS PLATTINENSIS FOERSTE, 1920;
CERAURUS CF. GLOBULOBATUS BRADLEY, 1930**

Upper Ordovician; Upper Bobcaygeon Formation; Ontario,
Canada; *right: Ceraurus plattinensis:* 7 cm; *left: Ceraurus cf.
globulobatus* (ventral): 3.3 cm

During more than a century of exploration and excavation within
its various outcrops, the Bobcaygeon Formation has produced
a wonderful array of cheirurid species, many of which are now
undergoing a scientific revision.

stars"—richly deserve both your attention and your admiration.

Among the more daunting challenges I faced while assembling *Travels with Trilobites* was deciding how to best organize the essays that appear here. Although all revolve around a solitary subject, they do cover an imposingly expansive territorial tract—both in terms of time and topography. During this process, the question was continually broached as to whether these fossil-fueled accounts would be most effective and enjoyable if

grouped by geographic region, by geologic period, or by related topic.

There were moments when it seemed that no solution would neatly encompass everything I hoped to present, and that no matter which editorial path was followed, not all content would fit succinctly into any such restrictive outline. Initially, instinct dictated observing a geographic plan, bundling the various featured trilobite locations by present-day continental alignment. I thought that such a strategy might potentially lend a degree of expansive "field guide" aura to the proceedings.

When push came to sedimentary shove, however, I decided on a timeline design, grouping the essays together by their respective geologic periods regardless of their corresponding location's current continental disposition. Thus, a report on the Cambrian trilobite outcrops of Siberia is directly followed by tales of similarly aged formations in Australia, China, California, Canada, and the Czech Republic. Chapters then proceed chronologically through the Ordovician, Silurian, Devonian, and Carboniferous until reaching a culmination with the demise of the trilobite line at the end of the Permian. If nothing else, such an approach provides a measure of comparative Paleozoic perspective regarding the ever-changing role trilobites played during this volatile period of Earth history.

Perhaps, when all is said and done, certain readers may conclude that writing about trilobites, to paraphrase the late Frank Zappa, is curiously akin to dancing about architecture. Maybe so. Indeed, the overriding philosophy espoused by the fossil-seeking field-workers featured within these pages has always been, "Just shut up, and dig!"

But whether or not you agree with this book's editorial focus, the insights presented, or even the choice of featured Paleozoic destinations, I sincerely hope you will find *Travels with Trilobites* to be an enlightening and entertaining journey into the Deep Time world inhabited by these most captivating of primal organisms. And if you happen to stumble upon an unintentionally erroneous factoid or two hidden within the text, photo captions, or specimen IDs, please take a moment to recall the immortal words of the old Scottish hard rock band Nazareth: In the end, it's all close enough for rock 'n' roll!

THE PHOTOGRAPHS

Much like the eternal question of whether a zebra is a white horse with black stripes, or a black horse with white stripes, I have long considered whether *Travels with Trilobites* is a photo book with accompanying essays, or an essay book with accompanying photographs. Either way, the avowed goal of this trilobite-propelled effort has been to achieve a somewhat harmonious balance between words and images, one that provides a proper forum for communicating the required degree of insight, fascination, and, yes, affection for those unique Paleozoic remnants.

The 300 photographs presented in *Travels with Trilobites* (which are of specimens drawn from my personal collection unless otherwise noted) have been specifically selected for a variety of reasons. Paramount among these is the simple desire to showcase the incredible diversity of design inherent within the trilobite class. This is something most effectively done when the featured trilobite is provided a full-page format from which to convey its oft-bizarre, half-billion-year-old elegance. Indeed, more than 50 such trilobite "portraits" embellish these pages, many representing unique examples being shown to the public for the first time.

From the graceful configurations displayed by Cambrian olenellids, to the streamlined shapes exhibited by Ordovician cheirurids, to the ornate intricacies of Silurian lichids, to the detailed compound eyes that adorn Devonian phacopids, my

intent is to furnish free rein to the expansive array of morphological forms that the trilobite line achieved during its 270-million-year trek through evolutionary time.

In addition, virtually every photo has been strategically placed either to best reflect the faunal content of each featured trilobite location or to enhance the particular paleontological concept being presented. Thus, an attractive grouping of Rochester Shale specimens has been aligned with the story of that deposit's famed Silurian quarry in upstate New York, just as a dramatic display of *Paradoxides* trilobites from all corners of the globe accompany the chapter on plate tectonics.

However, as has been proven on innumerable occasions since the dawn of trilobites, things don't always go exactly as planned—either with evolution or the layout of a book. As the various locations, files, and observations that comprise *Travels with Trilobites* were being finalized, an unexpected development began to rear its somewhat ugly cephalon. Despite the intended breadth and scope of this volume's editorial focus, it started to become abundantly clear that there simply weren't going to be enough categories in which to squeeze all the photographs originally planned for inclusion.

To remedy this situation, it was decided that at the end of every corresponding time-period chapter, a separate trilobite gallery would be created. Each of these five galleries (Cambrian, Ordovician, Silurian, Devonian, and Carboniferous) feature an assortment of age-appropriate photographs, compiled and presented without overriding regard to any specimen's location of discovery, regional recurrence, or scientific gravitas.

The photos appearing in each of these special gallery sections have not been chosen to reflect some storied paleontological principle, highlight the contents of some legendary quarry, or reveal some previously mysterious morphological trait. They're being displayed merely because of each trilobite's inherent beauty and stunning strangeness, qualities that make them more than worthy of being seen, studied, collected, and admired.

In the pages directly adjoining these various photo galleries, key "curation" information (identifying each specimen as to its genus, species, author, age, location, formation, and size), along with accompanying facts and field notes, has been provided for each specimen. Indeed, to the best of my abilities, this information has been supplied for *every* photograph in this book, whether it's of a relatively common *Asaphiscus wheeleri* or a rare as trilobite teeth *Uralichas hispanicus.*

Unfortunately, due to the "publish or perish" mantra that in recent years has seemingly come to dominate contemporary academic thought, trilobite nomenclature is in a near-constant state of flux. That's particularly true when it comes to both identifying and *mis*identifying the plethora of previously unknown species that continue to emerge from an ever-expanding inventory of new Paleozoic horizons. Thus, what may be considered cutting-edge trilobite data one week may be dismissed as little more than flawed, "old school" wisdom the next.

Oh, and by the way, the latest scientific research indicates, contrary to popular belief, that a zebra is, in fact, a black horse with white stripes. And when all is said and done, *Travels with Trilobites* is an essay book with accompanying photos. But those among us who may choose to rush directly to the photographs without paying much initial heed to the text are certainly forgiven. After all, we trilobite enthusiasts tend to be a very visually oriented lot.

METRIC MEASUREMENTS

You will quickly notice that all measurements presented in *Travels with Trilobites* are metric. That's true whether the size in question is indicating the dorsal length of a diminutive trinucleid trilobite or the distance between Paleozoic continents. Why? After all, being a lifelong resident of the United States, I am all too aware that in this corner of the planet such metric measurements are traditionally viewed with a pronounced level of disdain and dismissal, if not outright disgust. However, most trilobites were relatively small, and measuring these specimens in centimeters rather than inches has become an accepted international standard, one this trilobite-filled tome has somewhat reluctantly chosen to embrace. And once that initial metric threshold was breached, switching from feet to meters, and miles to kilometers, also seemed to make logical sense.

Such measurements may seem a little cumbersome at first, especially for anyone brought up under the auspices of the "renowned" American school system. For those who need a quick refresher, just remember that 1 inch equals 2.54 centimeters, and 1 mile equals 1.61 kilometers.

NEVADIA SP.

Lower Cambrian; Campito Formation, Montenegro Member; Esmeralda County, Nevada, United States; 5.1 cm

This is an elegant early trilobite from a remote and rarely collected location in the western United States.

(TOP, LEFT) CROTALOCEPHALUS AFF. AFRICANUS

Middle Devonian; Bou Tchrafine Formation; Jorf, Morocco; 12.3 cm

In recent years, Jorf has emerged as one of Morocco's premier Devonian trilobite locations.

(TOP, RIGHT) LENINGRADITES GRACIOSUS BALASHOVA, 1976

Middle Ordovician; Upper Llanvirnian and lower Llandeilian; Uhaku Regional Stage; Glyadino, St. Petersburg region, Russia; 7.2 cm

This beautifully preserved example of a relatively rare trilobite was drawn from one of the newer quarries that have recently emerged in the St. Petersburg region.

(MIDDLE, LEFT) DOLICHOHARPES DENTONI (BILLINGS, 1863)

Upper Ordovician; Lake Simcoe; Bobcaygeon Formation; Ontario, Canada; 4.5 cm

This is one of the more unusual species found within the abundant Bobcaygeon Formation layers.

(BOTTOM, LEFT) WANNERIA WALCOTTANA (WANNER,1901)

Lower Cambrian; Kinzers Formation; Lancaster County, Pennsylvania, United States; 6 cm

Once prevalent on the commercial market, these large trilobites have become scarce in recent years as area quarries continue to close. This represents one of the keystone species in any major trilobite collection.

(OPPOSITE PAGE) ELLIPTOCEPHALA SP.

Lower Cambrian; Rosella Formation, Atan Group; British Columbia, Canada; 8 cm

In the early years of this century, fossil-oriented adventurers began seeking out previously inaccessible locations . . . occasionally with spectacular results.

TRAVELS WITH TRILOBITES

INTRODUCTION

TRILOBITE MORPHOLOGY

Morphologically speaking, all trilobites appeared as variations on a surprisingly similar theme. Despite existing for 270 million years, and producing more than 25,000 scientifically recognized species, trilobites shared a strikingly uniform body plan—and there is good reason for these similarities. Even at a very early stage in the history of life on our planet, the trilobite design had already proven to possess a certain degree of evolutionary perfection.

Sure, not all trilobites looked the same. Some species had eyes; some didn't. Some trilobites reached humongous dimensions; others did not. Many possessed long, flowing, genal spines, whereas others displayed little more than vestigial stumps. A wide variety of species presented pygidia featuring ornate pleural furrows, and others exhibited telson-like tails that occasionally exceeded the length of their entire bodies. Some trilobites were configured like a hydrodynamic rocket ship, and others resembled nothing more than a primordial meatloaf. A few

ALBERTELLA LONGWELLI (PALMER AND HALLEY, 1979)

Middle Cambrian; Carrara Formation; Pahrump, Nevada, United States; 8.1 cm

This is a particularly eye-catching example of a rare western species. Recent advances in preparation techniques have made complete specimens of these thin-shelled trilobites more prevalent on the world market.

early species even highlighted a strange, multisegmented opisthothorax, an elongated anatomical feature that provides bold evidence of trilobites' even more primitive, wormlike predecessors.

Indeed, during their lengthy trek through time, trilobites existed in an almost dizzying array of shapes and sizes. Perhaps no other animal class in the entire history of Earth has displayed the diversity of design shown by these amazing arthropods. But at their heart (and, yes, trilobites apparently did possess a primitive but effective cardiovascular system), they were all remarkably alike.

Named not, as generally surmised, for their three main body segments—cephalon (head), thorax (body), and pygidium (tail)—but rather for the three lobes that longitudinally divided their dorsal exoskeleton, whether they were Cambrian olenellids, Ordovician asaphids, or Devonian phacopids, most trilobites presented a fundamentally analogous body plan. Such characteristics as librigena, interpleural furrows, anterior margins, and facial sutures (which aided trilobites in shedding their shells during a molt) were shared by a significant number of trilobite species, as were such exotic-sounding components as axial rings, articulating facets, and palpebral lobes.

Taken together, a listing of their structural features affords nothing less than a veritable thesaurus of tongue-twisting trilo-terms. But in their purest form, they also provide both the professional paleontologist and the amateur trilobite enthusiast with a convenient and effective means of comparing and contrasting the key anatomical elements presented by these ancient arthropods. By doing so, the majesty, mystery, and mystique shared by these animals begin to reveal themselves in all their Paleozoic glory.

Quite simply, their strikingly distinctive morphological configurations served to make trilobites totally unique life-forms. As superficially similar as some trilobites may appear to certain contemporary creatures—perhaps most notably, isopods and horseshoe crabs—any and all such similarities are nothing more than shell deep. Indeed, appearances aside, trilobites have no direct living relatives. They do share many of the key characteristics of the arthropod phylum with everything from brine shrimp to wood lice, but their primeval anatomy marks their members as a totally separate line on the planet's family tree.

So, when you look at the fossilized remains of a trilobite, whether it is a 3-centimeter-long Cambrian age *Elrathia kingii* or a 25-centimeter Ordovician *Isotelus maximus*, please consider that lurking under those imposing calcite carapaces were once animals that represented one of our planet's first, and most successful, experiments with complex life. From their initial moments on Earth some 521 million year ago, few animals were ever as evolutionarily "perfect" in their morphological form as that fascinating organism known as the trilobite.

THE FIRST TRILOBITES

Trilobites seemed to emerge fully formed upon the Cambrian scene. By the time, some 521 million years ago, the initial members of this ancient line of arthropods began filling marine environments around the globe, they were already creatures with highly developed eyes, complex digestive systems, and admirably functional calcite carapaces. At

***OLENELLUS FOWLERI* (PALMER, 1998)**
Lower Cambrian; Pioche Shale Formation; Nevada, United States; 6.5 cm

Note the long, multisegmented opisthothorax on this specimen, which has been preserved as an internal mold. Some trilobites from this site feature a black calcite shell.

FALLOTASPIS SP.

Lower Cambrian; Montenegro Division,
Campito Formation; Magruder
Mountain, Nevada, United States;
3.9 cm

Scientifically recognized as one of the
first trilobites ever to swim in Earth's
seas, examples of this genus have been
discovered in Morocco and Siberia as
well as in the western United States.

that moment in evolutionary history, trilobites represented the most advanced life-forms ever produced on Earth.

Due to their sudden and dramatic appearance in the Paleozoic seas, it is evident that trilobites evolved from earlier, more primitive organisms—creatures that left behind minimal evidence of their existence in the fossil record. Some of these ancestral forms may have more closely resembled segmented worms or articulated jellyfish than the distinctive three-lobed animals that we all recognize today.

At least a few of those possible Precambrian ancestors, such as *Spriggina floundersi*—found in the 550-million-year-old Ediacaran age rocks of Australia—appear to have possessed rudimentary body segments, and even basic genal spines. But despite such somewhat recognizable anatomical advances, these primeval organisms were a far cry from the trilobites of the Redlichiida order that some 30 million years later would dominate the world's seas.

The truth is that Earth's sedimentary stratum has guarded the secrets of the earliest trilobites—and their immediate predecessors—with a steadfast determination. Indeed, only a few locations across the face of the planet have so far revealed themselves to be the presumed cradles for these original experiments with complex animal life. In recent years, discoveries made in such trilobite-rich Lower Cambrian outposts as the Montenegro Formation of Nevada, the Sirius Passet Fauna of northernmost Greenland, and the Chengjiang Formation of Southern China have cast much needed fossiliferous light on this imperfectly understood period of Earth history.

Despite such findings, however, the hunt for the World's First Trilobite rages on. Currently, one of the prime candidates for yielding that initial member of the trilobite lineage lies in the limestone outcrops of the Sakha Republic in distant Siberia. There, localized Lower Cambrian species such as *Profallotaspis jakutensis* and *Bigotina bivallata* appear to be perched at the very base of the entire trilobite family tree.

In fact, some recent academic thought suggests that the calcite-clad trilobite class may have originally arisen in the waters that once enveloped these Siberian exposures. Over a relatively short period—perhaps less than 200,000 years—hundreds of divergent trilobite species may have subsequently radiated from that single Siberian locale to reach suitable marine environments across the face of the planet.

Yet not all scientists agree with this premise, and they have a substantial inventory of sedimentary-strewn data to support their skepticism. It seems that at almost the exact same time that these Siberian species were first leaving their mineralized marks upon the Paleozoic world, other early representatives of the trilobite line (including such genera as *Hupetina, Eofallotaspis, Serrania,* and *Fritzaspis*) were emerging in certain key biological hotspots around the globe—including in 521-million-year-old outcrops recently found in Spain, Greenland, Morocco, and the western United States.

Thus, academic thought concerning the possible birthplace of the trilobite lineage is undergoing near-constant updating and revision. As fresh fossil material is discovered and new analysis conducted, paleontologists have slowly begun to garner a more detailed understanding of the circumstances surrounding the emergence of trilobites in Earth's early oceans.

Amid a flurry of hypothetical conjecture, one current theory postulates that a key wellspring for early trilobite development may have been located on the oceanic shelf adjacent to the ancient microcontinent of Avalonia, the landmass that, in the earliest days of the Cambrian, rose as a volcanic arc along the northernmost border of the supercontinent called Gondwana. There, 521 million years ago, on some hospitable, shallow-water plateau located along the equatorial line, a recognized trilobite species may have reared its antennae-adorned cephalon for the very first time.

Exactly which environmental stimuli may have helped provide the *why* behind this inaugural wave of trilobite evolution is another issue still open to widely varying degrees of debate. But the Lower Cambrian development of key morphological features, such as a resilient outer shell—itself indicative of trilobites' ascension to major players in the drama surrounding complex life's first, tentative steps on our planet—were most likely the arthropod response to ever-growing threats within their marine habitat, which included everything from the emergence of more menacing predators to increasingly unpredictable climatic conditions.

However, even if we choose to theorize about when, where, or why the first trilobite appeared, or what their first genus might have been, there is no question that a mere instant later in geological time these extraordinary creatures were primed to begin the most fertile period of their entire history. Thousands of new species, including the likes of *Gabriellus kierorum*, *Bristolia bristolensis*, and *Olenellus fowleri*, would soon pervade the planet's Lower Cambrian seas and firmly establish a class of animals that would survive for the next 270 million years.

THE ORDERS OF TRILOBITES: ORDER FROM CHAOS

Trilobites exhibited an almost overwhelming variety of sizes, shapes, spines, and segments. Their body plans, although all following a fundamentally similar three-lobed symmetry, featured an incredible diversity of design. Some trilobites were imposing behemoths; others survived as barely there specks along the Paleozoic seafloor. Some had multifaceted eyes sitting atop towering stalks; others had no eyes at all. Some were covered in inch-long calcite quills; others possessed outer shells as smooth as the proverbial baby's bottom.

During their 270-million-year journey through evolutionary time, these incredible invertebrates generated more than 180 academically accepted families—an impressive number featuring nearly 5,000 genera and 25,000 recognized species. Such almost unimaginable longevity and multiplicity has continually presented paleontologists with an intimidating yet elemental challenge: how to best categorize and distinguish one group of trilobites from another.

Quite simply, the issue boils down to finding the most expeditious manner of classifying these primordial creatures, one that allows us to gain

***THALEOPS LATIAXIATA* (RAYMOND AND NARRAWAY, 1908)**

Upper Ordovician; Lindsay (Cobourg) Formation; Bowmanville, Ontario, Canada; 6.4 cm

This species reflects the incredible morphological variance attained within the 10 trilobite orders. With more than 25,000 scientifically recognized species, trilobites rank among the most diverse life-forms ever to emerge on our planet.

at least a basic understanding of which families, genera, and species should be grouped together, and which subsequently produced a logical line of descendants. Once we tackle this daunting dilemma, we can begin placing the resulting trilobites into taxonomic orders that are manageable

***PLATYLICHAS ROBUSTUS* GURICH, 1901**

Middle Ordovician; Ludibundus Limestone; Ludibund Stage; Wismar, Germany; 9.1 cm

Discovered inside a large nodule, this specimen was found partially exposed in loose matrix in the early years of the twenty-first century. Rather than being indigenous, it is possible that its host rock was transported to this locale by glacial movement.

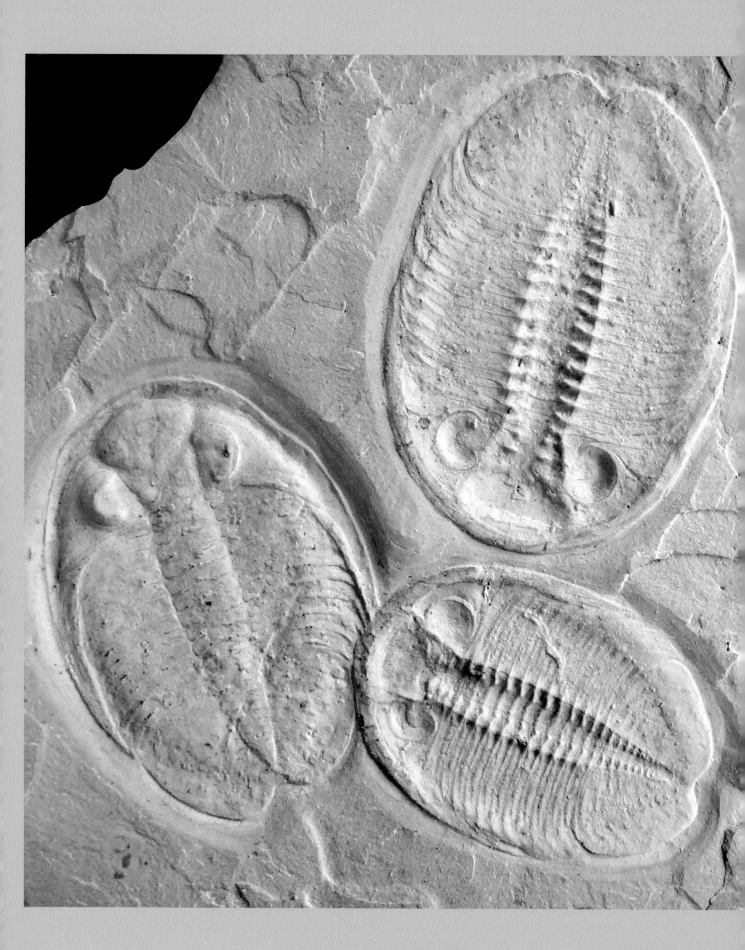

and to some extent practical. By doing so, scientists have tried to create a degree of mentally digestible "order" out of this arthropod-induced chaos.

After all, some trilobite orders, such as the Lichida, apparently arose in the Upper Cambrian and existed through the Ordovician and Silurian before ending their slither through the Paleozoic in the Devonian—a span of 180 million years. Then consider the Proetida; this order, consisting of thousands of species, produced members that first emerged in the Middle Cambrian and lasted all the way to the demise of the entire trilobite class at the end of the Permian, an impressive stretch of 250 million years. In dramatic contrast, the highly important early order Redlichiida arose, peaked, and vanished all within a roughly 20-million-year span of the Cambrian.

In all honesty, despite the best efforts put forth by the brightest minds in the paleontological field, there is still much debate when it comes to the topic of how to best align the trilobite class. Treatises have been published on the subject, and papers written, and then rewritten. Today there is still often controversy, if not outright contempt, surrounding some of the taxonomic classifications that have been handed down from the Mount Olympus of trilobite research.

Throughout the early years of the twenty-first century, it was generally accepted that there were nine distinct trilobite orders—the groupings into which every trilobite could be placed within some sort of basic evolutionary pattern. These orders were Proetida, Asaphida, Phacopida, Lichida, Ptychopariida, Harpetida, Corynexochida, Redlichiida, and Odontopleurida.

In 2020 a tenth order, Trinucleida, was proposed that would include certain species previously assigned to the Asaphida line. If that wasn't confusing enough, there is arguably an eleventh order, Agnostida, in which its uniformly diminutive species are continually being contested as being "true" trilobites by a growing number within the fossil fraternity.

Yet even these relatively well-defined classifications still find themselves constantly being tweaked and revised as new trilobite discoveries are made and additional information is garnered. And as more and more previously unknown genera emerge from freshly opened repositories around the globe, the task of identifying and classifying these recent arrivals has only served to add fuel to the Paleozoic fires surrounding what has already proven to be an incredibly complex and surprisingly challenging issue. Indeed, even within this distinctly invertebrate world, these far from made-to-order trilobite orders seem destined to serve as a major "bone" of contention for many years to come.

DALMANITES SP.

Silurian, late Llandoverian—early Wenlockian; Joliet Formation, Brandon Bridge Member; Waukesha County, Wisconsin, United States; largest trilobite: 4 cm

This is an interesting specimen showing both ventral and dorsal examples of a rare midwestern dalmanitid. Note the slightly inverted pygidium, a morphological characteristic that some believe assisted this species during enrollment.

THE CAMBRIAN PERIOD 1
541–485 Million Years Ago

SAKHA REPUBLIC, SIBERIA:
ROOTS OF A FAMILY TREE

SIBERIA. The name alone generates a variety of powerful images within the mind's eye, and few of them are particularly pleasant. Perpetually frozen landscapes and endless vistas of barren tundra—the outpost where wayward Soviet citizens were sent during the height of the Cold War, most never to be heard from again.

In fact, Siberia is all that we imagine—and more! Part of the Russian Empire since the seventeenth century, this immense tract of land covers more than 13 million square kilometers (the United States, by comparison, is barely half that size, and that includes the enormity of Alaska) and stretches from the Ural Mountains in the west, to the Arctic Ocean in the north, to the borders of Mongolia and China in the south, to the Pacific Ocean in the east. Even though it accounts for nearly 80 percent of Russia's total landmass, only a quarter of the nation's populace

BATHYURISCELLUS SINIENSIS IVANTSOV, 2005; *PHITOPHILASPIS PERGAMENA*

Lower Cambrian; Botomian Regional Stage; Sinsk Formation; Botoma River (right tributary of Lena River); Lena Pillars, Southern Yakutia; Sakha Republic, Russia; 12 cm (each specimen)

Phitophilaspis (lower) was a strange, trilobite-like arthropod that obviously—on the strength of this specimen alone—shared its marine habitat with early trilobite species such as *Bathyuriscellus*.

***BERGERONIELLUS GURARII* (SUVOROVA, 1956)**
Lower Cambrian, Series 2, Stage 4; Botomian Regional Stage;
Sinsk Formation near Lena Pillars (Lenskie Stolbi) National Park;
Lena River, Yakutia, Russia; 5.1 cm

In sharp contrast to Siberia's "desolate" reputation, the national park that borders this site features some of the most beautiful vistas in the Northern Hemisphere, at least during the area's short summer season.

calls this often inhospitable region home, with the net result being that Siberia ranks among the most sparsely inhabited places on the planet.

Siberia might not boast the requisite credentials to score high as a potential travel destination—although it does possess its share of breathtaking vistas, especially during the area's notoriously short summer—but this vast province has long earned high marks within paleontological circles, principally for its extensive Pleistocene fauna. Among the creatures found fossilized within the 14,000-year-old Siberian permafrost are mammoth, bison, cave bear, and a peculiar species of wooly rhinoceros that sported a sickle-shaped nose horn that frequently reached more than a meter in length. And recently, a perfectly preserved, mummified cave lion cub was discovered —with even its thick, tawny fur still intact.

Such finds have ignited hope that in the not-too-distant future science may devise a method for extracting enough viable DNA from the remains of these long lost animals to resurrect their various species. In anticipation of such a momentous event, an Arctic-adjacent expanse near Siberia's Kolyma River has already been transformed into a 16-square-kilometer nature reserve called Pleistocene Park, a facility expressly designed to incorporate these incredible creatures in their original habitat—if, and when, they happen to return from the past.

In addition to the headline grabbing notoriety generated by these Pleistocene discoveries, the rough and rugged Siberian landscape also houses a wide range of other fossil-filled sedimentary strata, some of which may prove to be of equal, or even greater, scientific import. In recent years, a series of key Paleozoic outcrops uncovered in the Sakha Republic—a remote district of the Russian Federation that is roughly the size of India—have begun to attract the attention of both paleontologists and amateur enthusiasts around the globe. The reason for such interest is easily explained; nestled among the locality's imposing limestone layers are the Pestrotsvet, and slightly younger Sinsk formations, 521-million-year-old Lower Cambrian storehouses that have produced some of the world's earliest representatives of the trilobite class.

As tantalizing as such primordial organisms may seem, simply reaching any of these Siberian trilobite strongholds requires more than the mere desire to do so. Such a distance-defying trek takes persistence, patience, and planning, along with an appetite for tackling some of the most formidable obstacles nature has at its disposal. In this secluded corner of the world, roads are often little more than centuries-old animal trails. Even specially designed four-wheel-drive vehicles frequently find themselves bogged down in the thick layer of mud that rises above the permafrost during the brief summer season when digging in this isolated part of Siberia—closer to Nome, Alaska, than to Moscow—is even possible.

Aside from the often unfavorable weather and the forbidding terrain, just arranging a trip to these nearly inaccessible Cambrian sites can present enough logistical headaches to make any right-minded soul second guess his or her original intent. Outsiders often require various visas, vaccinations, and credentials, and even Russian citizens can find organizing a proper Sakha Republic expedition difficult at best. Although the legendary Trans-Siberian Railway passes through this colossal land, it doesn't come close to contacting any of these fossiliferous locales. Unless someone knows just the right parties, and can grease just the right palms, acquiring the kind of properly equipped SUV or truck needed to undertake such an arduous adventure is a major procedural nightmare. Perhaps the only logical means of reaching these remote sites is by helicopter, with the nearest airfield being hundreds of kilometers away. By its very definition, however, the use of such a craft translates into both incredibly high costs and severe limitations in terms of both expedition members and available equipment, a combination that quickly negates any potential benefit.

Yet despite such apparent roadblocks, as is often the case with even the most far-flung trilobite destination, getting there can be half the fun. And once you've managed to reach one of these lonely Siberian outposts—which may take two or three days of travel, depending on your starting point and your primary means of transport—the rewards may very well justify any related expense, delay, or annoyance.

"A trilobite digging expedition to Siberia is limited right from the start," said Arkadiy Evdokimov, based in St. Petersburg, who runs the world's largest commercial outlet for Russian trilobites. "The best locations are more than 6,500 kilometers from St. Petersburg. It can require a series of flights, and then driving a very long distance over difficult terrain, merely to get close. We have done a great deal of research just to know where to go, because if you don't, you will waste your time. It is expensive and difficult. Every specimen you find is precious."

More than half-a-billion years ago things were far different in Siberia than they are today. In the Lower Cambrian, the area that now comprises the incessantly snow-swept steppes of the region's central plateau was positioned far nearer the planet's equator. What later became the Sakha Republic's fossil-bearing layers rested just offshore of what was then a large Siberian island-continent, a landmass located in the southern quadrant of the Panthalassic Ocean. It was apparently a very fertile marine environment, one in which scientists now believe some of the initial members of the trilobite line may have first reared their calcite-covered heads. Siberian species such as *Profallotaspis jakutensis* and *Bigotina bivallata* seem to appear at the very base of the entire trilobite family tree, animals that some 521 million years ago provided stunning proof that complex organic life could indeed survive on this "hostile" planet called Earth—especially if those primitive biological forms were encased in protective outer shells. Although early members of the trilobite lineage were almost simultaneously emerging in other biologic hotspots around the globe—as evidenced by primal Paleozoic outcrops recently discovered in Spain, Nevada, and Morocco—none has yet produced fossils that predate the material being unveiled within those distant Pestrotsvet Formation layers.

Despite some of the apparent morphological similarities exhibited by their respective trilobite faunas, not all of these early aquatic ecosystems shared a comparable climate in the Lower Cambrian. Although Siberia lay squarely within a tropical zone near the ancient equator, what would later become the Moroccan and Spanish outcrops were then found in southern seas with more moderate climates. Science has yet to present a comprehensive analysis of the role Cambrian ocean temperatures may have played in the development of the trilobite class. It seems virtually certain, however, that a hospitable climate—as well as a diversity of habitats within any given marine environment that would logically promote development of differing species—served a vital function in the successful emergence and eventual radiation of these ancient arthropods.

"It now appears that early trilobite species initially arose in tropical or temperate regions of the planet," said Bill Barker, an Arizona-based trilobite dealer and a noted authority. "There have been many new discoveries made around the world in recent years, though it does seem as if the Siberian examples represent the earliest yet found."

The Siberian search for the "World's Oldest Trilobite" has understandably garnered a cave lion's share of attention in certain paleontological circles, but the neighboring, slightly more recent Sinsk Formation also contains significant Lower Cambrian fauna. The Pestrotsvet Formation that produces *Profallotaspis*, *Bigotina* and their invertebrate ilk has continually proven to be among the most parsimonious of trilobite repositories—having yielded only a scarce few complete, articulated specimens to date—but the material in the Sinsk, although certainly not what anyone would label "bountiful," is clearly more abundantly distributed. Many of the trilobite examples found within these thickly banded limestone layers have come from outcrops that emerge along the closely aligned

Anabar and Lena rivers. These "sister" waterways bisect the heart of the Sakha Republic and cut a northerly route through the Central Siberian Plateau for nearly a thousand kilometers before emptying into the perpetually frigid waters of the Arctic Ocean.

There, in a variety of locations where the powerful river currents have exposed wide swaths of Sinsk Formation strata, complete—and often impressively large—examples of such trilobites as *Bergeroniellus asiaticus*, *Delgadella lenaica*, and *Jakutus primigenius* (some up to 15 centimeters in length) have been found. Almost always preserved with a thin, coffee-hued calcite coating that beautifully captures the subtle morphological variances displayed by each species, these distinctive trilobites have quickly become prime targets of acquisition for trilobite enthusiasts around the world—as well as being the subject of growing academic analysis. Amazingly, a substantial number of these 521-million-year-old species even display pronounced axial spines, perhaps the first time such an advanced anatomical feature is evident within the fossil record.

In addition to the wide-ranging commercial appeal and scientific importance of these Lower Cambrian species, trilobite enthusiasts should be excited to know that Middle Cambrian genera are also strongly represented within the Sakha Republic's fossiliferous layers. These examples can be uncovered in an equally remote sedimentary deposit, the 510-million-year-old Mayan Stage of the Anabar Plateau—a rugged spot of

JAKUTUS PRIMIGENIUS IVANTSOV, 2005
Lower Cambrian; Achchagy-Tuoydakh Lagerstatte; Southern Yakutia, Siberia; 11 cm

Found in Siberia's distant Sakha Republic, examples of this rare species have exhibited both axial nodes and pronounced axial spines. *Jakutus* represents one of the largest trilobite genera discovered at this locale.

permafrost-infused tundra located high above the Arctic Circle, some 40 kilometers west of the Lena River.

Although these Middle Cambrian trilobite sites are no easier to reach than those that hold their Lower Cambrian predecessors, the Anabar material discovered to date has at least been significantly more abundant. There have, in fact, been enough examples found of such diminutive species as *Hatangia scita*, *Michaspis librata*, and *Urjungaspis picta* that complete specimens of these generally 2- to 4-centimeter-long bugs have become readily available on the open market. More than a dozen different species have so far been unearthed from the plateau's various Mayan Stage outcrops, and the diversity of trilobite design found within these hard limestone sediments indicate just how rapidly—at least in geologic terms—the trilobite class was evolving at this still nascent time in its history.

"We have had a bit more luck finding the Middle Cambrian trilobites," Evdokimov said. "But they are still very difficult to discover. There are also many more partial specimens found than complete ones. But you try not to get too discouraged."

As an added incentive to seeking out these remote Paleozoic outposts hidden deep within the Sakha Republic, some of the area's richest fossil-laden exposures occur within the confines of the beautiful Lenskiye Stolby National Park. Created in 2018, this impressive refuge (larger than Wyoming's famed Yellowstone Park) has recently been designated a UNESCO World Heritage Site and ranks as one of the most topographically dramatic

OLENOIDES DUBIUS (LERMONTOVA, 1940)
Middle Cambrian, Amgan Stage; Lena River (basin); Siberia, Russia; 6.2 cm

This example is remarkably similar in appearance to *Olenoides* species found in Utah, China, and British Columbia.

destinations in the Northern Hemisphere. Among the park's premier features are the incredible Lena Pillars, 250-meter-high dolomitic cliffs—first formed by the ebb and flow of Middle Cambrian seas—that have been weathered over the past half billion years into striking geological formations that now tower over the winding waters of the Lena River. Unfortunately, as has already occurred in other parts of the world, the park's World Heritage designation may eventually serve to limit, if not totally halt, fossil exploration in the region—a fact that has made the area's recent trilobite discoveries even more precious to collectors around the globe. With its majestic rivers and massive pillars—along with its renowned diamond deposits, which annually produce over $600 million worth of precious stones—it's apparent that trilobites are far from the only natural treasures to be found in this isolated Siberian wilderness. However, the region's geologic "wealth" goes beyond even diamond extravagance and half-billion-year-old trilobites.

If one is searching for a singularly distinctive fossil genera within the area's Lower Cambrian outcrops, housed amid those same Sinsk Formation layers (occasionally in direct contact with such trilobite species as *Bathyuriscellus siniensis*) lurks a strange trilobite-like arthropod known as *Phytophilaspis pergamena*. Due to their atypical anatomical configuration, these remarkable creatures—believed to be a form of xandarellid—have commonly, if incorrectly, come to be referred to as "bilobites" by those who collect and market such specimens. Although these large, 10- to 15-centimeter-long animals bear a striking superficial resemblance to trilobites (and share such characteristics as prominent eyes, a large pygidium, and even the outline of their hypostome), they are also quite different due to the shape of their facial sutures, the reduced size of their thorax, and the fact that their limited number of thoracic pleura are fused to one another.

When all that we know is considered, however, perhaps it shouldn't be particularly surprising that the fossils of large, nontrilobite arthropods exist within these 521-million-year-old layers. After all, similar but decidedly different creatures appear in other Cambrian outcrops, most notably China's bountiful Chengjiang deposits and British Columbia's legendary Burgess Shale. Both of those fossil-filled formations are home to species of the legendary "trilobite eater" *Anomalocaris*, along with other imposing Paleozoic marine predators. And with our knowledge of life's earliest steps growing exponentially on an annual basis, it has become abundantly clear that whether one considers the fossils found in the sedimentary layers of China, Australia, British Columbia, or Siberia, the early Cambrian seas were virtually bursting with evolutionary energy. At no other time in our planet's 4.54-billion-year history have the forces of nature been more inventive or unpredictable.

"We are learning more about the Cambrian Explosion all the time," said Barker. "As recently as 2018, a major new Burgess Shale outcrop was discovered in western Canada, which featured some of the best-preserved specimens ever found. On top of that, Chengjiang and Emu Bay in Australia are still offering up surprises in terms of their arthropod content. And Siberia remains something of an understudied and underappreciated resource. Hopefully it will reveal more of its secrets in the years ahead."

For most of us, Siberia may forever hold a bleak image as one of our planet's true off the grid destinations, but this is a land rich in both abundant natural resources and plentiful paleontological reserves. Although it is highly unlikely that many reading this will one day find themselves ankle deep in tundra muck as they meander toward a summertime dig in the Sakha Republic's distant Pestrotsvet Formation, life could present far worse options. Indeed, merely considering the chance to explore for perhaps the oldest trilobites on Earth should stir the fossil-loving souls within each of us.

THE PALEOZOIC ERA

A mere listing of its geologic periods cannot begin to convey the singular significance of the Paleozoic Era. After four *billion* years of spinning through the cosmos as a virtually barren ball, Earth suddenly sprang to life at the dawning of the Paleozoic. It was a time that witnessed the most rapid development and diversification of multicellular organisms in global history—an event that signaled the beginning of the famed Cambrian Explosion.

The Paleozoic, a name that roughly translates into "the age of ancient life," lasted for 290 million years—from 541 to 251 million years ago. Science has chosen to divide that nearly incomprehensible span of time into seven geologic periods: Cambrian, Ordovician, Silurian, Devonian, Mississippian, Pennsylvanian, and Permian. Each of these periods not only featured its own unique flora and fauna but also apparently its own inherent natural disaster, which helped signal that period's end . . . and another's subsequent beginning.

Trilobites served to neatly bookend the Paleozoic, arising in the Lower Cambrian, some 521 million years ago, and lasting until the end of the Permian, more than a quarter-billion years later. But these amazing arthropods were far from the only creatures living in those ancient seas. In the Cambrian, a time when oceans dominated the planet and surrounded supercontinents like

GLOSSOPLEURA GIGANTEA (RESSER, 1939)
Middle Cambrian; Spence Shale; Antimony Canyon, Utah, United States; 11.4 cm

This specimen represents one of the largest Middle Cambrian trilobite species found in North America.

Laurentia and Amazonia, such trilobite species as *Olenellus transitans* and *Glossopleura gigantea* were joined by a plethora of other invertebrates. Many of these were nonbiomineralized creatures whose soft-bodied existence has been chronicled in varied locations across the face of the Earth, including paleontological treasure troves in China, Australia, and western Canada.

As dramatic evidence of the destructive powers that the planet routinely unleashed upon its new inhabitants, the Cambrian abruptly ended in the wake of the Earth's initial extinction event. During that episode, which may have been triggered by a pronounced drop in marine oxygen levels, 60 percent of extant ocean life perished—including many trilobite genera, in addition to species of brachiopods, conodonts, and soft-bodied arthropods.

By the dawning of the Ordovician, 485 million years ago, trilobites such as *Isotelus latus, Cybeloides girvanensis*, and *Amphilichas ottawaensis* still represented one of the world's dominant forms of life, but that dominance was beginning to be threatened by the emergence of primitive fish. By the end of that period, some sea creatures had become bold enough to venture permanently onto land, beginning to inhabit the fringes of the continental mass known as Gondwana. Yet disaster loomed again, and the planet's second major extinction event at the end of the Ordovician—possibly caused by a rapid global cooling—eradicated virtually all terrestrial life and severely depleted life in the seas as well.

The Silurian began 444 million years ago and witnessed a rejuvenation of life following this mass extinction. Trilobites, although in decline, still produced spectacular species like *Arctinurus boltoni* and continued to thrive in the world's oceans, where they were joined by an ever-growing variety of fish and various species of eurypterids. Life returned to the land, and plants began to dominate the global terrain. At this time, four distinct continents existed on Earth, providing ample new ecological environments in which diverse faunal forms could emerge and evolve.

By 419 million years ago, the Devonian was underway, and huge, armor-plated fish roamed the seas. These imposing creatures posed a further threat to the remaining trilobites, such as the heavily spinose *Drotops armatus*, as well as to virtually everything else that crossed their path. Plants continued to monopolize the land, helping to develop a more hospitable, oxygen-filled environment that allowed terrestrial life to flourish. But once again Earth proved a fickle host, and the planet's next great extinction served to eradicate 70 percent of all life—an event that emphatically signaled the end of the Devonian.

During the Mississippian and Pennsylvanian periods (commonly combined into the Carboniferous outside of North American paleontological circles), which began 359 million years ago, Earth's temperature rose dramatically, creating lush tropical swamps that pervaded nearly every landmass. It was the ideal environment for certain early plant species to flourish, which they did, eventually leaving behind the coal-rich deposits that provided the Carboniferous with its name. The seas were also particularly warm, often forcing such diminutive trilobite species as *Ameura major* and *Comptonaspis swallowi* to burrow into the surrounding ocean floor sediments for survival.

By the beginning of the Permian, 299 million years ago, all the Earth's landmasses had joined together to form the supercontinent called Pangea. The few remaining trilobites (all members of the Proetida order) were uniformly small—under 5 centimeters—and played a relatively insignificant role in the planet's continuing evolution . . . virtually all of which came to a crashing halt when the greatest mass extinction in Earth history wiped out 90 percent of life around the globe, subsequently putting a dramatic end to both the

Paleozoic Era and the trilobites' 270-million-year swim through time.

EMU BAY SHALE, AUSTRALIA: *WONDERS DOWN UNDER*

Kangaroo Island. It sounds like a locale drawn straight out of a classic adventure novel, a swash-buckling action flick, or at least a Disney theme park. It is an exciting name, one seemingly cloaked in mystery and intrigue, designed to make the mind's eye instantly wander toward visions of exotic ports and faraway destinations. In many ways, this remote outpost off the coast of South Australia more than lives up to its colorful moniker. Indeed, Kangaroo Island would be the ideal setting for any Hollywood blockbuster seeking rugged terrain, picturesque landscapes, and an abundance of wildlife along with a decidedly non-twenty-first-century ambiance.

Spanning slightly under 150 kilometers from tip to tip, and with a year-round human population barely topping 4,000—a number overwhelmed by the island's 60,000 resident kangaroos—KI, as the locals call it, has long been one of Australia's favorite vacation spots. Its seemingly endless stretches of sandy white beaches, dense Eucalyptus groves, and acclaimed nature preserves—featuring everything from seals and penguins to wallabies and echidnas—provide this oceanic oasis with a laid-back aura and a definite *National Geographic* vibe. There's even a thriving art community on the island—sort of a highbrow retreat for those seeking to escape the Big City "marsupial race" found in Sydney, Melbourne, or nearby Adelaide, which is situated 112 kilometers to the northwest. Clearly, Kangaroo Island has a rhythm, and a look, all its own. Lying only 13 kilometers off the Aussie mainland across the Investigator Strait, and connected by a frequently running ferry, in many ways this is a place far removed from the rest of Oz in terms of tone, temperament, and manner. And the fact is that KI has had quite a lengthy time during which to develop its extraordinary charms at its own unhurried pace.

First separated from the Australian continental mass around 9,000 years ago by a rise in surrounding sea levels, carefully constructed stone tools dating back 11,000 years have been found in a smattering of sites throughout the island. It is believed that all Aboriginal life ceased to exist on KI more than 2,000 years ago, but the reasons for this are still not fully understood. Some scientists suggest this disappearance may have been caused by sudden climate changes forcing all the island's native inhabitants to flee back to the mainland or be wiped out. Among many Aboriginal tribes now living on the Aussie mainland, KI is still rather dramatically known as Karta—the "Island of the Dead."

Aside from its rich history, natural beauty, and flourishing art scene, Kangaroo Island has risen to prominence in the minds of many, especially those who possess a decidedly paleontological perspective, for another reason—it is the site of one of the most important Lower Cambrian trilobite formations in the world. Along the island's north coast, among the weather-worn cliffs that outline the azure waters of Emu Bay, on an isthmus known as Cape D'Estaing, lies the Emu Bay Shale Formation. This dazzling location references KI's unique species of dwarf emu (*Dromaius novaehollandiae baudinanus*), the flightless bird that became extinct on the island during the early years of the nineteenth century due to a lethal combination of overhunting and habitat loss.

More significant, however, is that Emu Bay is one of the few places on the planet—along with such renowned Paleozoic troves as Canada's Burgess Shale and China's Chengjiang Formation—that presents a rich biota of soft-bodied Cambrian organisms intermingled among its hard-shelled

trilobite fauna, and some of those trilobites display perfectly preserved antennae and appendages. Since this remarkable site was first discovered in 1952, more than 50 distinct species—including trilobites, nonbiomineralized arthropods, sponges, brachiopods, and even a type of lobopod worm—have been unearthed amid Emu Bay's sienna-hued shale exposures. Scientists from both the Melbourne Museum and Adelaide's South Australian Museum still conduct occasional excavations at the site, and new species continue to pop up on a semiregular basis from within the shale's 10-meter-thick fossiliferous formations.

The deposit's 510-million-year-old trilobite specimens present an eye-catching appearance featuring subtle tones of yellow, orange, red, and brown. However, the coarse, granular texture of the formation's sandstone layers impedes the level of detailed preservation that often distinguishes the renowned Burgess soft-tissue discoveries, virtually all of which are fossilized in a finely grained mudstone. Still, such relatively minor geological "nuisances" have done little to detract from either the diversity of material emanating from the Emu Bay site or the timeless tale it tells.

"Nothing else can match Burgess," said Martin Shugar, a field associate in paleontology with the American Museum of Natural History. "But that's not to take away from the discoveries made at Emu Bay. The specimens found there are certainly impressive and important. Also, the quality of the Emu Bay rock doesn't seem to impact its trilobite preservation as much as it does its other fossils."

REDLICHIA REX HOLMES, PATERSON, AND GARCÍA-BELLIDO, 2019

Lower Cambrian, Middle Series 2; Botoman Equivalent Stage; Emu Bay Shale; Kangaroo Island, Australia; 15.5 cm

The various minerals found in the ground waters that pervade the Emu Bay Shale helped provide the rich display of colors seen here.

The formation's trilobites include 3- to 4-centimeter-long examples of *Estaingia bilobata* and *Holyoakia simpsoni*, along with large, often multicolored specimens of the outcrop's most abundant resident arthropods, *Redlichia takooensis* and the recently described *Redlichia rex*. These latter species occasionally reach impressive dimensions of more than 25 centimeters in length and closely resemble *Redlichia* types found in Chinese sedimentary layers of roughly the same age that now lie more than 7,500 kilometers to the north.

Few, if any, of these formidable trilobites fully retain the calcified remnants of their characteristic outer shells—with most being fossilized as detailed internal molds—but they are almost all beautifully preserved and usually feature a brightly tinged mineralized coating. Some of these imposing *Redlichia* specimens also display serrated predation scars on their carapaces. These jagged wounds indicate that despite their apparent "top trilobite" status in the area's Cambrian seas, they were not above attack from a throng of even more intimidating arthropods that lurked within this ever-hostile marine ecosystem.

At one time, especially during the final years of the twentieth century, magnificent, complete examples of Emu Bay trilobites were somewhat common sights at fossil shows around the globe. Indeed, for a brief period, displays brimming with large, colorful *Redlichia* specimens—presented by stereotypically easygoing and knowledgeable Aussie merchants—highlighted natural history extravaganzas held in Denver and Tucson. More recently, however, especially after the Australian government began to place stringent restrictions on commercial digging for fossils in and around the entire Cape D'Estaing vicinity, all material from that renowned repository has become incredibly rare—and incredibly coveted.

"Getting any fossils from Emu Bay is now nearly impossible," said Tom Kapitany, a leading

Australian natural history purveyor based in Victoria. "Anything I've seen over the last few decades has come strictly from old collections."

In contrast to the formation's comparatively prolific trilobite reserves, KI's soft-bodied material has always been exceedingly scarce. The fauna features evidence of such genera as the enigmatic wormlike *Palaeoscolex*, the strange ovate arthropod *Squamacula*, the so-called great appendage organism *Tanglangia*, and the nonbiomineralized trilobite clone *Kangacaris*. Of special note, the shale also presents fragmentary remains of the legendary "trilobite eater," *Anomalocaris*, the giant marine predator also known from partial fossils found in both Middle Cambrian Burgess and Lower Cambrian Chengjiang outcroppings. In 2011, what are believed to be the first known examples of stalked, multilensed *Anomaloracis* eyes were discovered in Emu Bay's Big Gully outcrop, indicating to academics that the intricate arthropod compound eye may have developed millions of years prior to the phylum's characteristically calcified exoskeleton.

In addition to the rarity, beauty, and scientific significance of these species, what the Emu Bay Shale is perhaps most renowned for among collectors are the highly unusual trilobites *Emuella polymera* and *Balcoracania dailyi*. Both are diminutive genera, usually less than 2 centimeters long, indigenous primarily to these formations as well as to a small, recently discovered outcrop in Antarctica. When found complete, these species can feature up to 90 segments within their elongated opisthothorax. A steadily increasing number of paleontologists have come to view this distinctive

ESTAINGIA BILOBATA (POCOCK, 1964)

Lower Cambrian, Middle Series 2; Botoman Equivalent Stage; Emu Bay Shale; Kangaroo Island, Australia; 2.2 cm

Most trilobite fossils represent fragments of molted exoskeleton rather than remnants of the deceased animal itself.

morphological feature—the extension of the trilobite's main body, culminating in a relatively tiny pygidium—as a probable holdover from the trilobite line's more primitive origins.

Some Lower Cambrian olenellid trilobites found in other areas of the planet (perhaps most notably *Olenellus fowleri* and *Bristolia bristolensis* from the western United States) also present a prominent, but comparatively smaller, opisthothorax. However, no trilobites found anywhere on Earth can rival either *Emuella* or *Balcoracania* in terms of their sheer segmented majesty, with their disproportionally elongated "tails" constituting nearly 50 percent of their total body mass.

"The trilobite material coming out of the Emu Bay Shale is truly special," said Kapitany. "Many of the species—especially *Emuella* and *Balcoracania*—are fascinating to both the collecting and scientific

REDLICHIA TAKOOENSIS (LU, 1950)

Lower Cambrian, Middle Series 2; Botoman Equivalent Stage; Emu Bay Shale; Kangaroo Island, Australia; 21 cm

This impressive example represents what is probably close to the upper size limit of this trilobite genus. A 2019 study separated *R. takooensis* from the new species *R. rex*.

BALCORACANIA DAILYI (POCOCK, 1970)
Lower Cambrian, Middle Series 2; Botoman Equivalent Stage;
Emu Bay Shale Formation; Cape D'Estaing, Kangaroo Island,
Australia; 2 cm

This early trilobite features up to 97 segments in its opisthothorax,
the most of any known species. Unfortunately, on this specimen
part of that distinctive morphological feature remains buried in
the surrounding matrix.

communities because they represent something so atypical."

Numerous Cambrian-age locations on Earth feature a more diverse array of trilobite species, but few Paleozoic epicenters have drawn more academic attention than the fossiliferous formations on Kangaroo Island. The reason for this interest is understandable; the sedimentary layers of the Emu Bay Shale are older than the 508-million-year-old Burgess Shale outcrops and younger than the 515-million-year-old Chengjiang Formation repositories. Thus, these 510-million-year-old Australian specimens provide a unique looking glass upon a primitive world where early life was still struggling to grasp a solid foothold, experimenting with countless strange shapes, sizes, and morphological configurations in its efforts to survive.

Comparably exciting paleontological stories are revealed in the fossils of the Burgess Shale, the legendary Middle Cambrian formation in British Columbia that has generated more scientific study—as well as often whimsical mainstream media analysis—than any other invertebrate locale in the world. Although the trilobite material emanating from Emu Bay seems most closely aligned with the appreciably less-renowned Chengjiang layers—with nearly identical *Redlichia* species being observed within both outcroppings—some soft-tissue examples unearthed on KI appear more analogous to the fossils drawn from the oft-studied Burgess biota.

The similarities, as well as the differences, between the specimens found in these various locations have drawn keen interest from academics who specialize in studying fossilized soft-body preservation. What particularly fascinates these paleontologists—beyond recognizing the incredible faunal diversity already present at this early, transitional stage of Earth history—is the concept that in the primordial past, Burgess and Chengjiang were both apparently cool, shoal-filled marine environments located adjacent to the continental shelf. In contrast, during the Cambrian, the area surrounding what is now the Emu Bay Shale was

CENTROPLEURA PHOENIX (ÖPIK, 1961)
Middle Cambrian, Series 3, Guzhangian; Boomerangian
Regional Stage; Upper Fauna; Christmas Hills Road, 11 km west of
Smithton, Tasmania, Australia; 12.4 cm

This is among the largest specimens ever found of this strangely
shaped trilobite. No complete specimen with full genal spines
has yet been recovered.

most likely a shallow, warm water, coastal plateau situated close to the equator.

That such apparently divergent Cambrian ecosystems managed to nurture a remarkably similar fauna—as well as present a habitat that would eventually allow for soft-tissue preservation—provides new and fascinating information to those studying this Paleozoic phenomenon. It has been a long-held hypothesis that only relatively deep-water shelf environments, where unexpected mudslides and sudden ridge collapses could quickly bury an area's invertebrate inhabitants, were suited for soft-tissue fossilization. Apparently, that is not the case.

As more potential answers are found for the riddles surrounding the initial emergence of complex life on our world, a corresponding number of new questions arise. Considering the still limited knowledge we possess of this pivotal time in Earth's development, such a situation is far from unexpected. As additional Lower Cambrian locations are discovered around the globe, it is hoped that they—along with the fossils found in sites like the Emu Bay Shale of distant Kangaroo Island—will provide some of the missing clues needed to further unravel the fascinating conundrum that is the early history of our planet.

SNOWBALL EARTH

It ranks among the most intriguing questions in the entire paleontological canon: What events triggered the famed Cambrian Explosion some 540 million years ago? Was that world-changing phenomenon caused by a meteorite strike that subsequently seeded the Earth's oceans with a flood of biomatter? Was it caused by the spewing forth of nutrient rich plumes from a field of undersea volcanoes, or by lightning bursts that electrified the planet's amino acid–laced atmosphere?

Was it perhaps the work of some Supreme Being's creative hand?

Over the years, each of these theories (and many more) has been presented to explain the sudden emergence of complex organic life on our world—a biological blooming that followed in the wake of four *billion* years of virtual inactivity. And some may even hold a loose grip on the proverbial smoking gun when it comes to properly answering such a momentous question. However, it was only during the latter days of the twentieth century that a particularly fascinating hypothesis began to take shape that offered the best explanation yet for the planetwide conditions that preceded this unprecedented outbreak of life. It was proposed that at some point significantly prior to the dawning of the Cambrian Period—possibly as far back as 700 million years ago—the entire globe had become entrapped within a layer of ice, a phenomenon that has since become popularized under the catchy moniker Snowball Earth.

Science has long known that our world has been subjected to fluctuating periods of glaciation and subsequent warming. To the best of our limited knowledge, however, none of these other ice ages featured a stage during which the entire planet had become covered in a thick coating of pole-to-pole permafrost. From the moment of its inception, the Snowball Earth theory had both ardent supporters and equally vociferous detractors.

Proponents quickly voiced their belief that such a premise would help explain the elevated sea levels that characterized the Cambrian, with the thawing of globe-blanketing ice sheets subsequently providing a more hospitable, and wide-ranging aquatic environment for the arthropod armies soon to follow. Those who questioned Snowball Earth countered that a complete glaciation of the planet, particularly a freeze that lasted for tens of millions of years, would have created an especially unforgiving global habitat. In such a world,

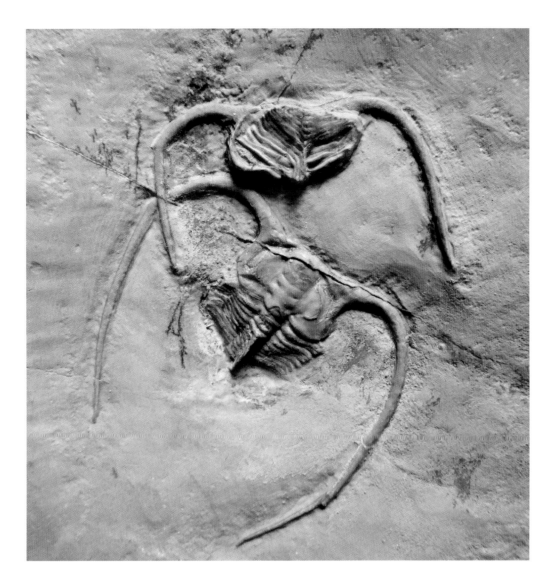

primitive Precambrian life-forms would have faced difficulty developing and surviving, let alone thriving.

Those who supported the Snowball Earth hypothesis insisted that marine volcanic activity—believed to be globally rampant both late in the Precambrian and throughout the Cambrian itself—would have eventually served to both counteract and slowly begin to reverse the more drastic aspects of such an ice world. In addition, they pointed out that these volcanic conditions might well have acted as the needed catalyst for the development of complex life in the primeval oceans. Indeed, even today we see a somewhat similar phenomenon happening on a much less dramatic scale in areas adjacent to undersea fumaroles.

Although we are now perhaps better able to address the question of what events may have triggered the Cambrian Explosion, the specific

HOLMIELLA FAIX
HOLLINGSWORTH, 2006

Lower Cambrian; Poleta Formation; Montezuma Range, Nevada, United States; 3 cm across genal spines

Here is a little-known and highly unusual early species hailing from the western United States. Both specimens may be displaying partial enrollment, perhaps the first known example of such behavior in the fossil record.

NEVADIA SP.
Lower Cambrian; Campito Formation, Montenegro Member;
Esmeralda County, Nevada, United States; 5.2 cm

This is an elegant example of the early trilobite line showing how anatomically advanced these arthropods were soon after their emergence in the Paleozoic seas.

natural causes involved with igniting that evolution revolution remain shrouded in secrecy. Certainly, with our current obsession concerning global climate change, we must also wonder if the atmospheric and geologic conditions required for the advent of an ice planet scenario could ever again rear their frosty fingers. Snowball Earth may have lasted for more than 100 million years, so it's safe to say that any return engagement would effectively end life—or at least *human* life—on our world. If such a disastrous scenario were to occur again, we could then compress the past 540 million years of Earth history into one short sentence: What the planet had giveth, the planet had taketh away.

CHENGJIANG BIOTA, CHINA: *MOUNTAINS AND MOLE HILLS*

Deep in the heart of the vast Asian continent, sequestered within the craggy cliffs, rolling hills, and tree-lined valleys that define much of China's Yunnan Province, lies the Lower Cambrian Chengjiang Biota, a series of 515-million-year-old mudstone formations that together constitute one of the most important fossil repositories in the world. In this traditionally isolated, generally impoverished, and culturally diverse region located within hailing distance of Vietnam's northern border, some of the oldest known trilobites and soft-bodied arthropods on Earth have been found. These enigmatic early Paleozoic relics predate the more renowned fossil material drawn from western Canada's historic Burgess Shale by more than seven million years.

From any imaginable perspective, these Chinese outcrops are filled with strange creatures, ancient "aliens" with intriguing names such as *Hallucigenia, Haikoucaris, Yunnanocephalus,* and *Microdictyon.* Some of these are armored worms—animals sporting a series of short appendages and no apparent eyes—and some represent early echinoderms, the phylum that eventually produced such familiar sea residents as starfish and sand dollars. Others are among the earliest representatives of the trilobite class, and still others are unlike anything else ever seen on this world, either before or since. Together these rare remnants of the distant past have begun to provide paleontologists with a special window through which to view the initial, tentative steps that life took on this planet during

EOREDLICHIA INTERMEDIA (LU, 1940)
Lower Cambrian; Chengjiang Maotianshan Shales; Heilinpu Formation; Yunnan Province, China; 9.2 cm

If you look closely, slight traces of antennae can be seen emanating from the cephalon of this large, colorful specimen.

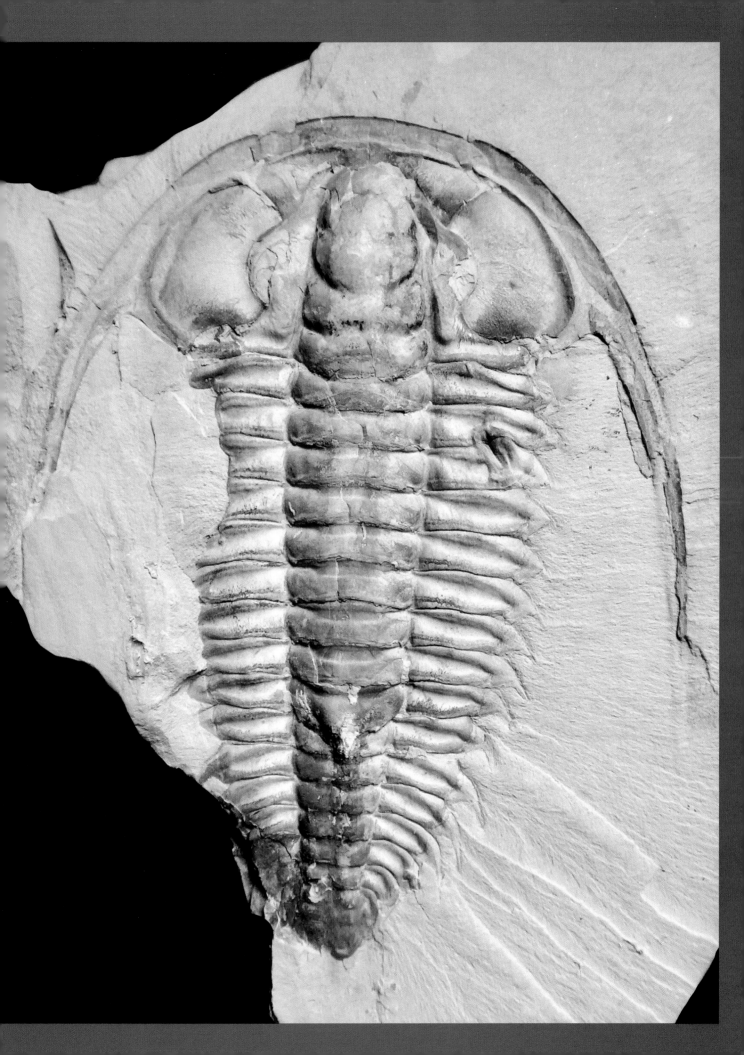

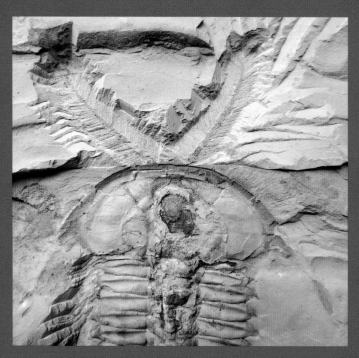

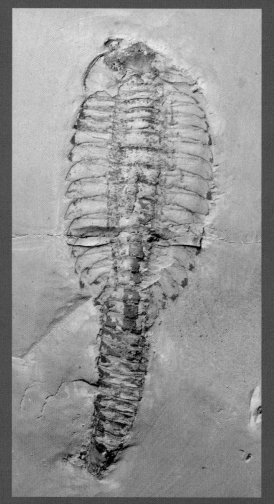

a time that has become popularly known as the Cambrian Explosion.

Since its discovery in 1984, the Chengjiang Biota has continually astounded the world's scientific community with both the quality and quantity of its distinctively well-preserved fossil fauna. In fact, many essential members of the site's disparate cast of soft-bodied animals are indigenous only to these layers—with what is now recognized as the Chiungchussu Formation producing more than 50 unique species alone. So far, 185 distinct organisms have been unearthed within the various sedimentary exposures that comprise Chengjiang's Maotianshan Shales, with more than half being

arthropods. Although virtually all apparently led to evolutionary dead ends—including, of course, the trilobites—a scant few possibly did not. With the site's formations presenting a superb array of sponges, worms, and jellyfish, in addition to its rich diversity of arthropods, the stratum represents one of the most prolific Lower Cambrian biotas ever found and, arguably, the most remarkable.

The shale's impressive arthropod assemblage includes four trilobite species—*Eoredlichia intermedia, Yunnanocephalus yunnanensis, Wutingaspis tingi*, and *Kuanyangia pustulosa*—all of which appear on the chalky, yellow-tinged mudstone as flattened, slightly calcified internal molds. Many of these specimens feature a mineralized patina that ranges in color from white to tan to brown, and over the past 30-plus years, these attractively toned trilobites, including complete examples of *Kuanyangia* and *Yunnanocephalus*, have been recovered in sufficient numbers to be spotlighted in major museum exhibits (as well as private collections) from Shanghai to San Francisco. Although trilobites comprise less than 1 percent of the Chengjiang Biota's abundant fauna, some of these 2- to 8-centimeter-long specimens have proven to be of particular note . . . especially those that feature fossilized antennae or appendages. In fact, a select number of the shale's *Eoredlichia intermedia* examples display unmistakable evidence of the earliest arthropod antennae so far revealed within the fossil record!

These exceptional trilobites also shared their Paleozoic domain with the globe-spanning "Terror of the Cambrian Seas," *Anomalocaris*, whose fossilized remains (mostly disarticulated grasping claws) have also been discovered in such varied locales as the Burgess Shale, Australia's Emu Bay Shale, and within at least two outcrops found in the western United States. And perhaps even more significant, among the Chengjiang Formation's stunning variety of fossils featuring nonbiomineralized,

(OPPOSITE PAGE, TOP LEFT) **KUANYANGIA PUSTULOSA (LU, 1941)**

Lower Cambrian; Chengjiang Maotianshan Shales; Heilinpu Formation; Haikou, Chengjiang County, Yunnan Province, China; 7.2 cm

This specimen represents one of the larger Chengjiang trilobite species. Such specimens are usually preserved as internal molds.

(OPPOSITE PAGE, TOP RIGHT) **HONGSHIYANASPIS YILIANGENSIS ZHANG AND LIN, 1980**

Lower Cambrian; Qiongzhusi Formation; Yunnan, China; 2 cm

Unfortunately, this specimen was damaged during the initial split-rock extraction, a common occurrence at this site.

(OPPOSITE PAGE, BOTTOM LEFT) **EOREDLICHIA INTERMEDIA (LU, 1940)**

Lower Cambrian; Chengjiang Maotianshan Shales; Heilinpu Formation; Yunnan Province, China; 3.8 cm

It is hard to miss the impressive antennae adorning this colorful specimen.

(OPPOSITE PAGE, BOTTOM RIGHT) **FUXIANHUIA PROTENSA HOU, 1987**

Lower Cambrian; Chengjiang Biota, Heilinpu Formation; Kunming, China; 6.3 cm

Although *not* a trilobite, this primitive soft-bodied arthropod shared the primal seas with their trilobite cousins.

Photo courtesy of HMNS

soft-tissue detail there is evidence of the first chordates. These include *Yunnanozoon lividum*, a tiny animal presenting what appears to be a primitive backbone. In some currently unimaginable way, this diminutive creature may represent the foundation of an evolutionary line that led *very* indirectly to the appearance of our own species more than 500 million years in the future.

"We've grown to understand a great deal about early life thanks to a century's worth of study of Burgess material," said David Rudkin, formerly of Toronto's Royal Ontario Museum. "But the arrival of fossils from other locations such as Chengjiang have supported some of our previous thoughts, while turning others upside down."

The Chengjiang Biota is situated nearly 2,500 kilometers southwest of China's cultural, political, and spiritual center in Beijing, yet much of Yunnan Province—which is slightly larger than Germany—might as well be a million kilometers away in terms of its grasp on the twenty-first century. The often oppressive reach of China's central government rarely touches Yunnan, and for the province's 47 million inhabitants—especially those living anywhere other than the colorful, bustling capital city of Kunming, which accommodates over seven million inhabitants, or the heavily industrialized town of Qujing, home to six million—life in this distinctly rural region remains little changed from a century ago.

In these rustic outskirts, a major means of transportation is still by oxcart, and for many locals such "conveniences" as reliable electricity and indoor plumbing are as foreign as an iPad. However, despite the backward appearance that some parts of Yunnan Province can convey to Western eyes, the area's residents are uniformly astute as well as surprisingly curious about the outside world. Rather than battling against any perceived threat posed by modern technology, when given the opportunity, those who live in and around Chengjiang (population 150,000) do their best to embrace it. Even when Chinese geologists first visited the Yunnan region in the late 1970s looking for an efficient means of tapping into the vicinity's rich phosphate deposits (used primarily as an agricultural fertilizer), the interest shown by locals was immediate and irrepressible. After all, since at least the tenth century, their families have been living in the nearby hills, utilizing its resources, as well as occasionally encountering strange "petrified" shapes in the rocks.

As it happened, by sheer good fortune, those phosphate deposits lay directly above the thick, fossil-bearing layers of the Maotianshan Shales. The subsequent mineral mining somewhat unintentionally triggered the discovery of the area's now renowned Lower Cambrian formations, along with their inherent, arthropod-rich fauna. Without great prompting, or even the direct promise of financial gain, the province's residents have always been eager to impart information about their homeland to travelers visiting this bucolic territory. In the process of doing so, they've equally sought to absorb any new knowledge that might be provided by these roving bands of tourists and academics.

In recent decades the Chengjiang locals have yearned to understand one thing more than anything else: What was significant enough about the soaring cliffs that have surrounded them their entire lives to draw legions of scientists to their remote corner of the globe? They had witnessed caravans of mining machinery enter their region and seen truckloads of phosphate carted out. They had observed teams of government officials wander through their hills, probing and testing for other mineral and fossil riches. They had even seen their friends and families head off to work for the various geological operations. Yet it wasn't until Chinese paleontologist Hou Xian-guang began uncovering, examining, and cataloging the amazing content of the Chengjiang Biota in the mid-1980s that the area residents—as well as the world's academic community—began to comprehend the answer to that question.

Almost immediately after the first fossil-oriented endeavors had begun in the area, these residents started to understand that something special had been found in their hills. Perhaps most didn't fully fathom that the remains of organisms more than half-a-billion years old were being excavated in the nearby sedimentary strata and that these discoveries would lead to a greater understanding of Earth's distant past. Even fewer would have grasped the concept that some of those finds would eventually make their way to the instant international marketplace provided by eBay, where rare Chengjiang arthropods and trilobites could sell for hundreds or even thousands of dollars. But such facts seemed to matter little in terms of either the level of interest or the subsequent labor these locals supplied. Their efforts, directed toward finding, extracting, and then breaking apart layer upon layer of the Lower Cambrian rock, led to the eventual recovery of thousands of fossil specimens, many of which soon formed the Yunnan Geological Survey's core collection.

However, as is too often the case with large-scale scientific ventures in underdeveloped parts of the world, there is another side to this story—one that questions where exploration ends and exploitation begins. The imposing, 50-meter-thick strata of mustard-colored mudstone that comprises the three principal formations of the Maotianshan Shales may have initially appeared impervious to anything outside of an army of giant bulldozers and modern rock-moving machinery. Yet since serious paleontological excavations began in 1984, an "army" of another sort—hordes of local residents spurred on by a heady mix of curiosity and the chance to earn a small wage by breaking rock and finding Lower Cambrian fossils—has begun to turn the mountains that surround Chengjiang into something more resembling mole hills. Hundreds of metric tons of rock have been removed by hand, but the damage done to the area's topography has never been properly offset by the minimal benefits to the local economy from the fossil and mineral trade.

After nearly two decades of uncontrolled digging at the site, both for fossils and phosphate, at the end of the twentieth century the Chinese government suddenly began to realize the inherent ecological dangers posed by the ongoing work being conducted in and around Chengjiang. The government took immediate, drastic, and somewhat uncharacteristic measures to protect the locale, as well as its invaluable early Paleozoic fauna. In 2001, they placed the entirety of the Maotianshan Shales under the jurisdiction of a National Historical Park, which immediately limited access to the site, and in 2012 the location was provided with a World Natural Heritage designation, which was intended to strictly control all future excavations conducted throughout the vicinity.

"Something needed to be done, or the entire formation could have been decimated," said Sam Stubbs, an associate at the Houston Museum of Natural Science. "The Chinese government needed to intercede, and they did."

Whether or not any of us choose to involve ourselves with the political machinations that will soon seemingly control all fossil-related activities conducted within China's expansive borders—where the export of all fossil material, whether it be dinosaur, pliosaur, crinoid, or trilobite, has now essentially been banned—the scientific importance of the material being found in Yunnan Province's Lower Cambrian outcrops remains beyond question or debate. When all is said and done, the Chengjiang Biota may well prove to be the most ancient site in the world featuring such an impressive range of early Paleozoic fauna. Indeed, it is difficult to imagine a more primal or revealing Cambrian strata being uncovered anywhere across the face of the planet.

The latest academic data indicate that more than 40 Cambrian-age locations around the globe have revealed themselves to be cradles of early multicellular life. These are so-called Goldilocks Zones—geological layers where the fossil evidence is of just the right age and state of preservation to provide a unique view of what primitive life was like in the oceans soon after complex biological organisms began their perilous trek through evolutionary time. Some of these Cambrian sites provide only tantalizing hints as to what these early creatures may have looked like—usually disarticulated pieces of trilobites with their hard, calcite-covered shells, well-developed, crescent-shaped eyes and impressive variety of species. But other localities reveal much more. In these particularly illuminating formations, not only have the remnants of complete trilobites been preserved in the finely grained sedimentary deposits but so have the fossilized impressions of a wide array of soft-bodied creatures. And when these astounding half-billion-year-old life-forms are placed within their proper Paleozoic context, they begin to shed a revelatory light on the Deep Time past.

The most famous of these captured-in-time repositories is unquestionably the 508-million-year-old Middle Cambrian Burgess Shale of British Columbia, perhaps the most studied, lauded, and legendary invertebrate outcropping on Earth. A similar, slightly older Cambrian biota featuring both primitive soft-bodied fossils and an assortment of early trilobites can be found in the 510-million-year-old Emu Bay Shale off the coast of South Australia. Here, animals comparable to those found at the Burgess site appear to mingle with many totally unique life-forms.

Despite the legendary status of Burgess and the current paleontological renown being showered upon the material emanating from Emu Bay, a provocative yet fundamental concept has slowly but surely begun to be embraced by the academic community: the sedimentary layers being extracted from the fossil-rich hills of Chengjiang contain the earliest look we've yet had at the amazingly diverse fauna that first filled our primitive seas. Although there may be older trilobites (known from 521-million-year-old rocks found in Siberia, Morocco, and Nevada), so far only the 515-million-year-old Maotianshan Shales have provided such a stunning perspective on the degree of biological heterogeneity already present on our planet soon after the dawn of the Cambrian Explosion.

When all the information is carefully considered, we are left to wonder why this unprecedented eruption of multicellular life occurred at this particular moment in our shared global history. Theories involving an increase in marine oxygen levels, a spike in undersea hydrothermal activity, as well as au-courant postulations concerning a Snowball Earth have all been presented by leading scientists in their attempts to explain why our world suddenly burst forth with life in near perfect synchronicity with the dawning of the Paleozoic.

Locations such as Chengjiang, Burgess, and Emu Bay each present their own fossiliferous clues as to why some of the most important changes in Earth's history took place over a relatively short period of geologic time, but the simple fact is that we may never know the full answer to this most basic of life-confirming questions. However, as more amazing discoveries are made in the rich Maotianshan Shales of southern China, and as additional scientific analysis is conducted on those finds, we are steadily moving closer to drawing back the curtain and revealing at least some of the secrets that surround the emergence of early life on our world.

CAMBRIAN TIME MACHINE

For the sake of this discussion, let's say you are the Ultimate Explorer, someone who has somehow gained access to a fully functional time machine. Thus equipped, you are able to travel back to any point in the past—from the birth of Earth some 4.54 billion years ago to your romantic dinner date last Wednesday. With the entire history of the world quite literally at your fingertips, perhaps your interests would lead you to return to the Old West to enjoy the sight of buffalo by the millions traversing the Great Plains. Or maybe you'd journey to the time of the Roman Empire to experience Julius Caesar at the peak of his oratory powers. Perhaps your curiosity would propel you back to the savannas of eastern Africa two million years ago to witness the lifestyle of our earliest hominid ancestors.

But maybe you'd choose to be *really* adventurous. As you sit at the cushy controls of your time machine and prepare to press the "go" button, perhaps you would plug in a date some 500 million years in the past. Such a bold commitment would certainly require plenty of initiative, planning, and preparation, for the world you would encounter upon arrival would be far different than the one you had just left behind. Indeed, the environs of Middle Cambrian Earth would resemble those of an alien planet, which in all honesty is exactly what this primal realm would be.

If your time machine happened to emerge on the recently formed continent of Laurentia, you would immediately note a virtually barren landscape. Amid active volcanic fields and frequent earthquakes caused by ever-shifting continental masses, you would see no animal life. The primitive flora you might encounter would resemble little more than microbial mats dotting the edges of rock-strewn tidal basins. The average daytime temperature might hover near an idyllic 70° Fahrenheit (21° Celsius), but the surrounding, acrid-smelling atmosphere would present a profusion of survival problems for any modern human—carbon dioxide counts nearly 16 times higher than twenty-first-century levels and an oxygen content only 60 percent of what we enjoy today.

In your preparations for this half-billion-year jump back in time, maybe you were resourceful enough to bring along diving gear that would allow you to explore the primeval seas. If so, you'd be perfectly prepared to encounter quite a different world where life would appear abundant, varied, and nothing less than astonishing. Bizarre invertebrate creatures of all sizes and shapes would fill virtually every marine niche. Amid patches of filtered sunlight, legions of small, spinose arthropods would float by on gentle ocean currents, and a horde of trilobite species such as *Olenoides superbus*, *Amecephalus althea*, and *Marjumia typa* would either jet past in water-propelled bursts or scamper across the seafloor on rows of spindly appendages. If you were fortunate, you might even glimpse a large, predatory *Anomalocaris canadensis* as it menacingly glided along—just don't risk putting any fingers near its razor-sharp mouth plate!

As you traveled through those Cambrian seas, a surprising reality would quickly become evident; in contrast to long-held scientific thought, trilobites did not dominate these waters, at least not in numbers. Five hundred million years ago veritable legions of soft-bodied animals inhabited the same offshore sanctuaries, among them an assortment of arthropod species that apparently outnumbered their trilobite relatives by a significant margin. Judging by fossil evidence found in paleontological hotbeds where just the right kind of Cambrian fauna has been preserved—including British Columbia's Burgess Shale, China's Chengjiang Formation, and Australia's Emu Bay

OLENOIDES SP.

Middle Cambrian; Wheeler Formation; Utah, United States; 12.6 cm

Olenoides represent one of the key genera of the Middle Cambrian. Transitional examples such as this—bearing morphological similarities to both O. *nevadensis* and O. *trispinus*—can occasionally be found amid the rocks that line the canyons of Utah.

Shale—the ratio of these soft-bodied creatures to trilobites may have been as high as 20:1.

Despite these seemingly lopsided numbers, trilobites like *Asaphiscus wheeleri* and *Kochina vestita* held a major evolutionary advantage over such delicately designed organisms as the "lace crab" *Marrella splendens* or the bizarre "velvet worm" *Hallucigenia sparsa*. Quite simply, trilobites possessed a hard, calcite-coated outer shell, and the multitude of soft-bodied animals often surrounding them in those early seas did not. Thus, while trilobites were better equipped to fend off

KOCHINA VESTITA (RESSER, 1939)

Middle Cambrian, Miaolingian Series, Wuliuan Stage; Langston Formation, Spence Shale Member; Box Elder and Cache counties, Utah, United States; 8 cm

Among the lesser known of Utah's Middle Cambrian species, examples of this midsized trilobite, which attained sizes up to 15 centimeters, have been unearthed in colors ranging from black, to brown, to red.

attacks by predators, these prolific nonbiomineralized morsels may have suffered the dire fate of serving as little more than brine-seasoned bouillabaisse for the larger, stronger inhabitants of their domain.

It is somewhat surprising, considering the apparent faunal abundance that filled the Middle Cambrian seas, that life in those oceans was already in pronounced decline, especially when compared to the maritime world that had blossomed soon after the advent of the Cambrian Explosion, some 20 million years earlier. A conspicuous drop in speciation had already begun to impact the trilobites, and as hydrogen sulfide levels caused by undersea methane eruptions began to rise and marine oxygen levels continued to dissipate, survival for these early oceanic inhabitants became more tenuous.

All things considered, any time-traveling excursion back to the Middle Cambrian seas would be a trip characterized by amazing sights and unexpected perils. Yet even knowing the dangers we might be required to face, are there many among us who would pass on the opportunity to participate in such a monumental journey? After all, even a brief sojourn back some 500 million years would provide a unique opportunity to witness complex life at one of its earliest and most important stages. Such a mission would offer unparalleled insight into an intricate ecosystem that has long held paramount scientific significance. We would be visiting an aquatic domain filled with time-honored secrets, one in which many key questions surrounding our planet's first arthropod inhabitants have only recently begun to be answered.

Despite a spate of fresh discoveries that have shed valuable new light on this enigmatic period of Earth history, innumerable unknowns surround the incredibly diverse fauna that inhabited that Cambrian world. And some of those secrets are likely to remain—at least until some intrepid explorer in the future, perhaps with the help of a time machine, stumbles upon the key to unravel the myriad mysteries that surround one of our planet's most enduring stories.

LATHAM SHALE, CALIFORNIA: *TRILOBITE GOLD*

The scenic vistas are incomparable. Every gentle curve in the road presents another postcard perfect view of sky, sun, and soil joining to form a colorful homage to the best that Mother Nature can offer. Driving in air-conditioned comfort along legendary Route 66 toward southern California's majestic Marble Mountains is an experience designed to equally stir the mind and the soul. As the sun rises over these jagged, weather-beaten peaks, its rays occasionally catch a crystalline outcrop, causing this impressive yet desolate landscape to sparkle like a precious jewel. In the minds of many, that's exactly what this rugged stretch located near the heart of the unforgiving Mojave Desert is—one of the most magnificent spots on the face of the entire North American continent.

Despite its imposing grandeur, to label the environs in and around this mountain range as anything other than "inhospitable" would do injustice to the word. It's hot. It's dry. At times it's downright brutal. During peak summer months, midday temperatures will often reach 120° Fahrenheit (49° Celsius), and legend has it that eggs opened on any dark surface will cook through in less than a minute. The surrounding terrain possesses its own awe-inspiring splendor but often appears as barren as an alien wasteland, with the high altitude and boulder-strewn topography broken only by the occasional appearance of a sturdy California barrel cactus.

Here in the grasp of the desert, cars frequently overheat, as do any unwary occupants trapped within those vehicles. The surrounding ring of towering peaks, which occasionally reach over 2,000 meters into the perpetually cloudless heavens, seems to further confine and magnify the sun's relentless intensity. It's all enough to create a potentially lethal environment for anyone brave—or foolhardy—enough to test the limits of this unforgiving tract of terra firma.

The inherent beauty, as well as the inherent danger, of this often challenging territory has been part of the American mindset for more than a century. Some 150 years ago, when the Wild West was still at its rip-roaring apex, those who chose to trek through the Marble Mountains on their meandering path northward to California's lush central farmlands did so with either an impressive sense of determination or little true awareness of the hardships that lay before them. Many of the concerns that faced those early pioneers, especially the unrelenting heat and the corresponding scarcity of water, still serve as essential components of local lore. Stories of courageous nineteenth-century travelers who lost their lives to the ruthless elements are stark reminders to anyone today who chooses to venture into this parched enclave in search of rock-climbing thrills or off-road adventures.

Unquestionably, this is a demanding land, one of America's last great wildernesses. Located just a four-hour car ride due east of the bright lights of Los Angeles, plenty of hearty souls are eager to partake in the Marble Mountains' myriad natural charms. Among those who flock to these picturesque mountains on a regular basis—usually during the region's far more temperate fall and spring—are a particularly dedicated group of visitors whose primary interest lies not in enjoying the range's rustic beauty or participating in the available extreme sports activities. These intrepid trailblazers focus on pursuing something perhaps even more extraordinary—the Paleozoic bounty housed within the Marble Mountains' sedimentary strata.

BRISTOLIA BRISTOLENSIS (RESSER, 1928)
Lower Cambrian; Latham Shale Formation; Marble Mountains, California, United States; 10.1 cm

Due to its rarity and beauty, this ranks among the world's most sought after trilobites. Recent digs in both the Latham and neighboring Pioche shale formations have yielded a smattering of complete specimens.

At the core of these outcrops lies the 518-million-year-old Latham Shale, source for some of the most spectacular and important Lower Cambrian trilobites to be found in the western United States. Featuring an impressive array of closely aligned species, including *Olenellus fremonti*, *Olenellus clarki*, and *Bristolia bristolensis*, the trilobites of the Latham Shale serve as exceptional examples of early life's first flowering on our world. Usually 5- to 10-centimeters in length (but on occasion considerably larger), and wonderfully preserved in a thin brown calcite "wafer" that contrasts dramatically against the toffee-hued limestone matrix, these organisms emerged less than three million years after the trilobite line first appeared on Earth, a remarkable development in the planet's ongoing evolution.

The rarity, beauty, and scientific importance of these trilobites have long provided impetus for fossil-loving folks from around the globe to venture into this dazzling yet often daunting desert habitat. Throughout the years, hundreds of complete trilobite specimens—highlighting a score of distinct species—have been discovered within the Latham Shale's abundant layers, and many of those examples now fill shelves in both museum displays and private collections. For many North American trilobite enthusiasts, an *Olenellus* culled from the Marble Mountains has come to symbolize the Lower Cambrian more than any other single fossil, making an archetypal specimen

of that genus a veritable "must" for any significant Paleozoic display.

The olenellid trilobites that emerge from these rocks represent the midstage of an incredibly successful family of ancient arthropods, with the fossilized remains of their various genera now being found everywhere from California to Pennsylvania to Scotland. Yet whether due to some still unknown limitations within their morphological design, or the shifting climatic and oceanic conditions that dominated this often volatile stage of Earth's history, the olenellids only appear in the fossil record for approximately five million years, a mere instant in geologic time. Thus, the trilobites of the Latham Shale provide a special glimpse at the faunal changes that were occurring in the world's oceans more than half-a-billion years ago, a time when multicellular life was first beginning to win its struggle for survival in the primal seas.

"The trilobites that come out of the Latham Shale are certainly among the most attractive and important Lower Cambrian specimens to be found anywhere," said Richie Kurkewicz, a California-based digger and collector of Cambrian species. "The Latham Shale has been worked for many years, and the results of those efforts have given science a wonderful chance to see how early life was flourishing and changing, especially regarding the trilobites."

A variety of other fossil-bearing Lower Cambrian layers are housed within the Marble Mountains, including the 515-million-year-old Chambless and Cadiz formations—home to such highly collectible trilobite species as *Nephrolenellus geniculatus* and *Mexicella robusta*. Yet unquestionably, the most renowned and studied geological component of this region is the 15-meter-thick exposure of Latham Shale that appears sporadically along the western slopes of the Providence Range, within which the Marble

BRISTOLIA INSOLENS (RESSER, 1928)
Lower Cambrian; Latham Shale Formation; San Bernardino County, California, United States; 7.3 cm

Examples of this early trilobite species are distinguished by their uniquely angled genal spines. Disarticulated head shields are common finds throughout the Latham Shale strata, but complete specimens are exceedingly rare.

Mountains are a primary member. In these layers, the easily fractured matrix produces fossilized trilobite parts in impressive numbers, and in some Latham horizons disarticulated *Olenellus* heads of all sizes can appear on virtually every other extracted block. The discovery of complete trilobites, especially examples representing any of the five recognized *Bristolia* species, is still cause for a major celebration by any hammer-wielding adventurer fortunate enough to uncover one. And though the hunt for trilobites continues to be the clear focus of most collectors' energies, a healthy assortment of other faunal elements can be found throughout the Latham, including abundant brachiopods and primitive sponges, along with the occasional highly coveted echinoderm.

There has been a healthy degree of scientific speculation regarding why fragmentary trilobite remains so dominate the Latham Shale layers. Some academics suggest that these deposits may represent a 518-million-year-old trilobite molting ground, a place where these intriguing invertebrates gathered on a regular basis to shed their protective calcite shells. Others have postulated that the powerful offshore currents of the Panthalassic Ocean, which then surrounded the ancient supercontinent of Laurentia and in which the trilobites that now fill the Latham Shale once congregated amid a series of shallow offshore reefs, served to tear apart fragile trilobite exoskeletons soon after the animal's demise.

"If you contrast the faunal content of the Latham Shale with that of the Pioche Formation of Nevada, another outstanding Lower Cambrian site, the trilobite preservation is considerably different," said Kurkewicz. "For whatever reason, there are more complete trilobites found in Pioche, while Latham produces more fragmentary specimens."

Unlike many other prominent western trilobite localities, such as British Columbia's renowned Burgess Shale or Utah's popular Wheeler Formation, hunting for Latham Shale fossils amid the majestic mountains of San Bernardino County has a rather obscure and relatively recent history. Due to the intense climatic conditions that too often dominate the region, it is easy to understand why some scientists were initially hesitant to explore this harsh land, and why others still prefer to let amateur "weekend warriors" take on the abundance of digging responsibilities, as well as the corresponding risks.

A variety of Native American tribes—including the Yuma, Serrano, and Cahuilla—have lived for centuries in proximity to the Mojave Desert. And from time to time since the late 1800s, small bands of settlers have attempted to carve out a meager living amid the region's rugged environs. But it wasn't until the early years of the twentieth century that southern California's prolific Paleozoic formations started to draw more than a modicum of scientific attention. It was then that Nelson Horatio Darton of the United States Geological Survey first began investigating the Providence Range's Cambrian exposures—including those of the Latham Shale. In the process of doing so, he mapped the region's most prominent sedimentary outcrops and gathered together samples of representative fauna, a collection that featured an assortment of disarticulated trilobite fragments drawn from the Latham's rich layers.

After conducting some preliminary studies on his finds, Darton turned over the bulk of his research, along with the corresponding specimens, to the legendary Charles Walcott. At that time, Walcott was in his initial year as the secretary of the Smithsonian Institution and was two years away from his monumental Middle Cambrian discoveries at British Columbia's Burgess Shale. His

recent work in Utah had helped transform Walcott into the world's foremost authority on Cambrian fossils, a title he accepted with considerable relish. Ironically, at first glance Walcott incorrectly identified the Latham material as being Middle Cambrian. Such an age would have made it contiguous with his Utah discoveries, as well as with what would soon become his pet project, the Burgess Shale, which contained fossils that were, in fact, 10 million years more recent.

OLENELLUS CLARKI
(RESSER, 1928)

Lower Cambrian; Latham Shale; Marble Mountains, California, United States; 6 cm

This is a sleek example of perhaps the most common trilobite found within the Latham Shale.

OLENELLUS CF. TRANSITANS
(WALCOTT, 1910)

Lower Cambrian; Echo Shale Member;
Nopah Range, California, United
States; 7.4 cm

This is a dramatic example of a classic
518-million-year-old species. It can be
distinguished from other olenellids by
its more streamlined body shape and
the position of its glabella.

In late 1907, Darton decided to take matters back into his own hands. He wrote and subsequently published a small paper in the *Journal of Geology* that outlined his Latham Shale discoveries and clearly identified the material as being Lower Cambrian. Perhaps due to the fragmentary nature of his finds, at that time his efforts generated only minimal interest within the scientific community. It wasn't until after World War II, when teams of academics and collectors started to make their own pilgrimages to the Marble Mountains in search of trilobites, that the site, and its corresponding fauna, began to receive due credit as a significant paleontological resource.

In the ensuing decades, a trip to the Marbles became nothing less than a bucket list necessity for trilobite enthusiasts who considered themselves true "players" in this timeless Paleozoic quest. Especially during the 1980s and 1990s—the height of the region's trilo-mania—collectors from all parts of the world were known to plan extensive, and often expensive, family vacations to this sacred site in the rugged American West merely for the thrill of trying to uncover their own *Olenellus clarki* or *Bristolia insolens*.

Despite the almost magnetic lure these Lower Cambrian layers have long held for those interested in trilobites, the Latham Shale is about to enter a challenging new chapter in its storied history. Due to the advent of recent legislation, much of the fossil-bearing matrix in the Marble Mountains is currently situated within the confines of what has been designated as the Mojave Trails National Monument. This places virtually the entire Latham Shale outcrop inside a federal park controlled by the Bureau of Land Management—an agency within the Department of the Interior—a fact that greatly restricts the rock-breaking ambitions of anyone in pursuit of Paleozoic prizes. Much of the area is no longer open to digging by unauthorized visitors, and in recent days even those sporting haughty museum permits or spotless university credentials have been turned back, or at least delayed, in their efforts to further explore this hallowed ground.

We can forever argue whether these restrictions, well-intentioned or not, truly benefit the preservation of natural resources or whether they are merely an unwanted and unnecessary infringement on citizens' rights to use public lands. No matter which side one may take in this debate, there is a growing trend throughout North America (and the world) to issue voluminous government rulings designed to protect fossils, minerals, and distinctive geologic formations that are perceived, or at least categorized, as "national treasures."

A full complement of trilobite material has long since been culled from the Latham Shale's Lower Cambrian layers—more than enough to furnish both scientists and fossil enthusiasts with an abundance of opportunities to continue learning about this unique Paleozoic outcrop. So even if collectors are no longer encouraged to break apart these primal shale layers, California's rugged Marble Mountains remain a worthy, if challenging, destination for anyone interested in uncovering evidence of life's initial efforts to establish a tenuous hold in Earth's ancient oceans.

TRILOBITES ON LAND

It was a perfect spring evening along Florida's east coast. With cloudless skies and an ambient temperature of 74° Fahrenheit (23° Celsius), an after-dinner stroll down one of the Sunshine State's legendary beaches was clearly in order. The walk would be illuminated by a fast-rising full moon that cast this deserted stretch of sea and sand in a sedate shade of blue. Rather than being abandoned, however, the beach proved to be packed—not by the sunburned streams of suburban *Homo sapiens* that had filled it to capacity just hours earlier but by scores of wriggling *Limulus polyphemus*, horseshoe crabs.

At some basic but essential level, these creatures *looked* so much like trilobites that this unexpected scene got me thinking. Despite horseshoe crabs and trilobites sharing the same taxonomic phylum, those ancient arthropods left behind no direct descendants. Yet, I wondered,

could it be that on occasion trilobites *acted* like horseshoe crabs and that at certain key moments in their life cycle they too ventured onto land? The mere concept seemed almost preposterous, if not borderline blasphemous. But apparently it is also true.

Despite the long-standing, well-established notion that trilobites were strictly and solely marine inhabitants, in 2014 scientists investigating a series of Lower Cambrian deposits in the Appalachian Mountains of Tennessee discovered something they had not anticipated. In outcrops of the 515-million-year-old Rome Formation, a team led by the University of Saskatchewan found unmistakable fossil evidence indicating that soon after beginning their lengthy crawl through Deep Time some olenellid trilobites had already started journeying onto dry land—or at least onto adjacent tidal flats. The early Paleozoic rocks emerging from that expedition revealed two distinct types of trace fossils: cracks and tracks. Cracks were indicative of a shallow seafloor habitat that began to rapidly bake and fracture upon exposure to direct sunlight. Tracks, however, told a decidedly more dramatic story. These particular trilobite trackways, also known as *Cruziana rugosa*, presented telltale signs that more than half-a-billion years ago those highly adaptable animals were exploring the surrounding shoreline and taking advantage of the resource filled tidal flats that bordered their aquatic home.

Considering the evident ability of these creatures to adjust to both environmental conditions and ecological opportunities—which they did continually throughout their 270-million-year history—it seems logical that even in the nascent stages of their quest for survival some trilobite species had found shallow-water marine ecosystems to their liking. From there, it would be just a matter of evolutionary expediency for these trilobed invertebrates to occasionally march their morphologically advanced body designs a few extra meters toward the surf's edge . . . and possibly beyond!

As long as trilobites prevented their gills from drying out, they could have partaken in sundry seashore activities while functioning outside their protective oceanic cocoon. After all, nature had furnished them with hard calcite shells, and thus equipped they may have been perfectly prepared to be the first animals on Earth to temporarily venture out from under the planet's shielding blue waves. In retrospect, perhaps such a revelation shouldn't come as a total surprise to those intimately familiar with the workings of the arthropod line. After all, horseshoe crabs are routinely found out of water during times of spawning. At full moons, especially in spring months, these "living fossils" emerge from the briny depths to deposit their eggs along marine perimeters. Many trilobites apparently exhibited (and perhaps initiated) behavioral patterns akin to those displayed by modern arthropods, so such Cambrian conduct—involving spawning, feeding, molting, or mating—is certainly not out of the realm of Paleozoic possibility.

The sum total of this time-tested, trilobite-centric information provides a major boost to the growing scientific speculation that terrestrial lifeforms may have initially evolved from marine animals. It had long been academically accepted that species of advanced organic life first appeared in freshwater pools, or at least in brackish intertidal

OLENELLUS ROMENSIS (RESSER AND HOWELL, 1938)

Lower Cambrian; Rome Formation (Montevallo Shale); Montevallo, Alabama, United States; 6.9 cm

These Rome Formation *Olenellus* may represent a species that first ventured out of the sea and onto adjacent tidal flats.

bays, prior to launching their primary land invasion, but now such attitudes are being called into serious question. Despite the limited fossil evidence supporting any claim that trilobites helped bridge this momentous sea-to-land transition, they may have in fact played a consequential role in pioneering a process that led to the eventual emergence of terrestrial creatures throughout our home world.

BURGESS SHALE, BRITISH COLUMBIA, CANADA: *CRADLE OF EARLY LIFE*

Some 508 million years ago, on a shallow offshore plateau located in what is now the Rocky Mountains of British Columbia, Canada, a legion of strange, almost other worldly animals ate, swam, procreated, and did just about everything within their limited powers merely to survive. Possessing neither the protective outer shells of their trilobite cousins, nor the size, strength, or smarts required to ensure their continued existence, life may well have been short and not particularly sweet for many of these primitive, sea-dwelling inhabitants. One can imagine the tenuous times of these unimposing arthropods, playing hide-and-seek with predators as they darted in and out of ancestral sponge beds and under sturdy stromatolite reefs in their attempts to simply make it through one more day. Some made it; some didn't. Those that managed to survive attacks by the mud-dwelling carnivorous worm known as *Ottoia prolifica*, the predacious trilobite *Olenoides serratus*, or the terrorizing giant arthropod called *Anomalocaris canadensis* still may have suffered a disastrous fate. The fossil record tells us so.

The Middle Cambrian was a time of great instability on a fast-changing planet. Underground volcanic eruptions, methane explosions, and major earthquake activity may well have been near

constant facts of life for the inhabitants of the Earth's primeval seas. The threat of suboceanic seismic events and habitat destroying mudslides may have presented a far greater challenge to both these apparently fragile, soft-tissue creatures and their sturdier trilobite relatives than all the era's predators put together. Alas, how exasperating it must have been to spend virtually all of your short existence escaping from razor-fanged sea monsters only to be done in by a rampaging pile of mud.

In retrospect, however, that mud was far more than a mere inconvenience to the local arthropod population. Without it our knowledge of the planet's primitive seas—and of its amazingly diverse invertebrate inhabitants—would be severely compromised. Current scientific thought postulates that a preponderance of the nonbiomineralized, soft-bodied specimens that we now know of through the fossilization process were at one time or another buried by these mudslides—underwater avalanches that quickly covered everything in their vicinity. By doing their dirty work so quickly and efficiently, these slides served as nature's time capsules, engulfing nearly the complete ecosystem within their path of destruction. In sites like western Canada's famed Burgess Shale, virtually every other overturned sedimentary rock presents some evidence of ancestral life—perhaps a small, soft-bodied arthropod or an early brachiopod. More often a disarticulated remnant of a trilobite species is found, such as *Kootenia burgessensis* or *Ogygopsis klotzi*, some with their delicate appendages still preserved.

OLENOIDES SERRATUS (ROMINGER, 1887)
Middle Cambrian; Burgess Shale Formation, Campsite Cliff Shale Member; Mt. Stephen Trilobite Beds; Field, British Columbia, Canada; 6.4 cm

This is an attractive specimen found in the 1960s within the legendary trilobite layers that sit atop Mt. Stephen.

Photo courtesy of G. Lee Collection

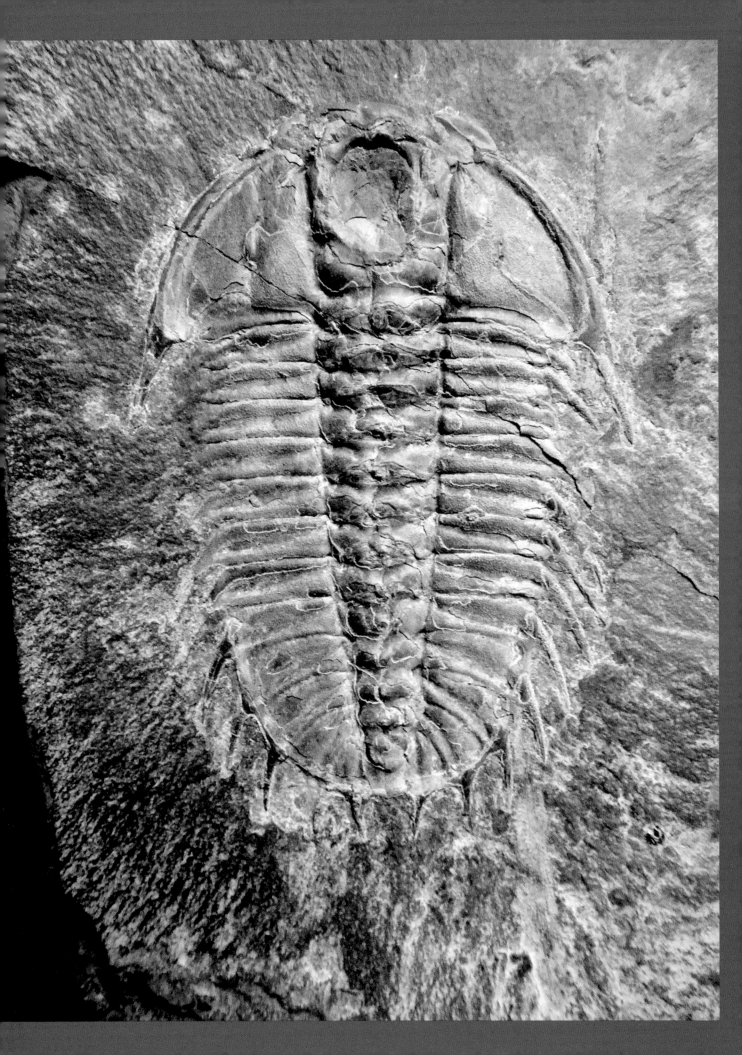

Occasionally, this natural burial process was accomplished with a minimal degree of damage to the organisms trapped within that menacing mudslide's suffocating grasp. Over the ensuing eons, layer after layer of sediment slowly formed atop these captured-in-time life-forms, and as the geological pressure escalated, these creatures—ranging in size from a centimeter to perhaps 18 centimeters in length, and except for the trilobites possessing no hard, calcite shell to provide three-dimensional substance—were slowly transformed. As tens of millions of years passed, these small, soft-bodied animals were compressed into little more than razor-thin films of calcium-laden aluminosilicate, or delicately detailed blotches of mineral-rich phosphate, both of which eventually left distinctive telltale stains on the surrounding shale surface.

How fortunate for science—and for anyone fascinated by the Earth's primitive past—that such accidents of nature occurred. Without them, the jigsaw puzzle that is our planet's earliest history would be lacking some of its most critical edge pieces. Indeed, some of these Burgess Shale soft-bodied specimens now rank among the most revered, studied, and famous fossils in the entire history of paleontology.

"Soft-bodied specimens like the ones you find in the Burgess Shale provide us with a spectacular and, to some extent, unexpected look at early ocean life-forms," said Utah-based trilobite enthusiast Terry Abbott. "The odds of any soft-bodied

OGYGOPSIS KLOTZI (ROMINGER, 1887)

Middle Cambrian; Burgess Shale Formation, Campsite Cliff Member; Field, British Columbia, Canada; 7.6 cm

This specimen has both of its free cheeks firmly in place, an unusual occurrence at this site where most of the recovered specimens are molts.

creature becoming fossilized is extremely remote, and that is why such specimens, especially those from Burgess, have become so well-known. They are without question among the most important invertebrate fossils ever discovered."

It isn't unreasonable to assume that anyone even remotely familiar with the fossil realm possesses at least some awareness of the Burgess Shale's renowned fauna. After all, the sedimentary layers contained within this half-billion-year-old outcrop of the Stephen Formation are home to the most celebrated collection of trilobites and soft-bodied animals in the world. Since it was first noted

OGYGOPSIS KLOTZI (ROMINGER, 1887)

Middle Cambrian; Burgess Shale Formation, Campsite Cliff Member; Field, British Columbia, Canada; largest specimen: 8.5 cm

This is perhaps the most prevalent trilobite species found within the Burgess layers. Hundreds of complete examples have been recovered although most are molts lacking free cheeks.

Photo courtesy of G. Lee Collection

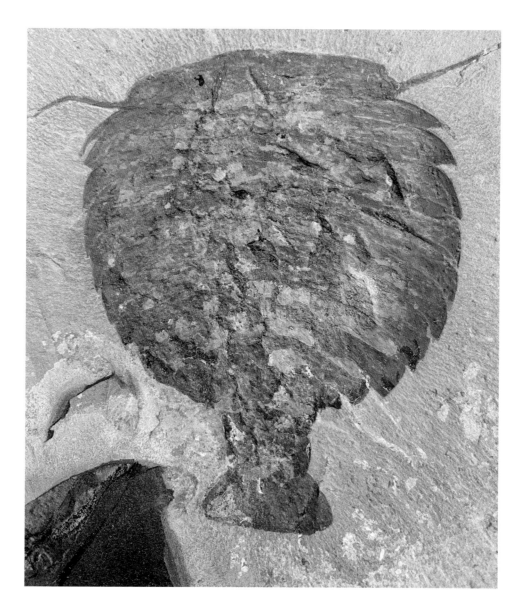

SIDNEYIA INEXPECTANS (WALCOTT, 1911)

Middle Cambrian; Burgess Shale Formation, Campsite Cliff Shale Member; Mt. Stephen Trilobite Beds; Field, British Columbia, Canada; 14 cm

Related to trilobites, this arthropod represents one of the Burgess fauna's most notable soft-bodied species. The matrix surrounding this large, impressive example has been chemically lightened.

Photo courtesy of The Natural Canvas

in 1886 by operatives working for the Geological Survey of Canada (in association with the construction of the Trans-Canada Railway), this site in the Canadian Rockies, located 2,280 meters atop Mt. Stephen, has enjoyed more mainstream focus and academic analysis than any other invertebrate locale ever found.

Much of that interest can be directly attributed to the celebrated work of the legendary Charles Walcott, who first discovered the location's fossil-bearing layers in August 1909 after months of searching for the shale's thinly banded exposures. Over the ensuing 15 years, this ubiquitous explorer returned to break rock at the site each summer, frequently with his family in tow. He often named new discoveries after his children, including the large arthropod *Sidneyia inexpectans*, a species first noted by his eldest son, Sidney, on the opening day of the 1910 digging season. At that time, finds at the Burgess outcrop were so plentiful that Walcott contentedly noted how a large knapsack full of carefully wrapped fossils could be acquired simply by digging steadily from sunup to sundown. After these annual summer pilgrimages, Walcott would further study this Paleozoic bounty and document his finds in a variety of outlets, including a wildly popular Burgess Shale report that appeared in the June 1911 edition of *National Geographic.*

During his years in British Columbia, Walcott uncovered literally thousands of Middle Cambrian fossil specimens, including both trilobites and soft-bodied arthropods, the majority of which are now housed in Washington, D.C.'s Smithsonian Institution. It was Walcott's other responsibilities at the Smithsonian—where he served as secretary from 1907 until his death in 1927—that kept him away from further research at the Burgess quarry and led somewhat indirectly to the 1924 excavations of the site by Harvard's Percy Raymond. It has been estimated that the combined efforts of the Walcott and Raymond digs produced well over 100,000 fossil specimens, virtually all of which featured Paleozoic species never previously seen by the eyes of science, or anyone else.

As is too often the case with historically or scientifically important specimens (think of the famous final scene in the initial *Indiana Jones* movie), for the next four decades these finds languished unstudied and virtually forgotten in a back room at the Smithsonian. It wasn't until another Harvard paleontologist, Harry Whittington, undertook his own Burgess expedition in the late 1960s that fresh attention began to be paid to these phenomenal discoveries. In the wake of Whittington's efforts, the first detailed scientific papers focused on the extraordinary Middle Cambrian fossils of the Stephen Formation were published in 1975. These scholarly documents confirmed that much of the Burgess Biota consisted of previously unknown organisms, many of which represented entirely new phyla. Even the often-studied representatives of the trilobite class were shown to exhibit hitherto unseen soft-tissue preservation, including antennae and walking legs, which further enhanced both the appeal and the accessibility of these ancient arthropods.

This groundbreaking research thrust these long-gone Burgess Shale inhabitants squarely into the public spotlight. Detailed, microscopically enhanced photos of the fauna's soft-bodied residents revealed a wide assortment of remarkable morphological features—gills attached to legs, multiple randomly placed eyes, and probelike pincers perched on the end of elongated snouts—that astounded academics and laypeople alike. Due to this influx of imagination-sparking material, along with the incredible story it told, in the latter decades of the twentieth century the Burgess site became the target of a veritable media storm. The shale's diverse trilobite species and diminutive soft-bodied sea-dwellers reemerged on the world scene as nothing less than the planet's original "rock stars." Their outlandish forms were featured everywhere from *Time* magazine covers to tomes that extolled the site's natural abundance (highlighted by the American biologist Stephen Jay Gould's award-winning 1989 book, *Wonderful Life*), to lushly produced television specials focused on evolution's first tentative efforts in the primal oceans. Each endeavor further cemented the Burgess Shale's reputation as one of the most hallowed and significant paleontological outposts on Earth.

Even in the twenty-first century, ongoing Burgess excavations led by Desmond Collins of Toronto's Royal Ontario Museum have continued to draw an astonishing degree of international interest. During the short summer digging season (which traditionally runs from late June through mid-September), hundreds of tourists travel to this legendary locale from around the globe to partake in guided tours of the site. They do so despite being well aware that "outsider" digging is strictly prohibited and that any unauthorized personnel who so much as touch a piece of loose scree with anything more than their boot clad foot run the risk of receiving a gentle but firm rebuke from one of the accompanying guides.

According to Collins, many of these visitors undertake the arduous, 22-kilometer, 11-hour round-trip hike—and pay the $126 per-person fee—both to experience a close encounter with one of the planet's most important paleontological repositories and to be able to tell their friends and families that they once stood at the site of Charles Walcott's renowned Burgess Shale dig. Although most quarry visitors display at least a modicum of fossiliferous interest, others— especially those lodging at the magnificent, century-old Fairmont Chateau Hotel perched along nearby Lake Louise—seem to take part in the stimulating journey up Mt. Stephen simply to get some exercise and enjoy the unmatched scenery. For these adventurers, the mountain's renowned fossil fauna is just an end-of-the-trail bonus that happens to accompany the world-class view. Indeed, the surrounding snow-capped environs are among the most breathtaking in the world; all the more reason that the challenging trek to the famed Walcott quarry has become a heart-pumping rite of passage for fossil enthusiasts from Shanghai to San Francisco.

"The quarry view is unmatched, and the fauna one finds there is nothing less than remarkable," said Martin Shugar, a field associate with the American Museum of Natural History. "The soft-bodied fossils, along with the trilobites found at the Burgess site, tell us of an amazing time in Earth's past. This period of the planet's development just abounds with fascinating questions, some of which we are now just beginning to answer."

It would almost appear as if their enigmatic origins represent a proud calling card for Burgess's legion of soft-tissue specimens, many of which were initially classified as "trilobitomorpha" in the historic 1959 *Treatise on Invertebrate Paleontology*. Perhaps no other paleontological finds have generated more debate, speculation, and outright awe than these blatantly bizarre life-forms.

As strange and wondrous as dinosaurs may seem, at least they possessed some anatomical characteristics to which we can relate—things like skin, bones, and teeth. Even trilobites, for all their quirky charms, presented a basic body design that was somewhat analogous to that of a few modern (although not directly related) animals, such as isopods or horseshoe crabs. But some of these soft-bodied creatures were really *out there*! One particularly whimsical Burgess species seemingly employed the use of no less than *six* stalked eyes, and other genera apparently "walked" on tentacles growing along their backs.

The Burgess Shale has produced an unrivaled trove of miraculously preserved soft-bodied specimens—along with abundant supplies of such trilobite species as *Kootenia burgessensis*, *Ehmaniella waptaensis*, and the previously mentioned *Olenoides serratus*—all of which have helped provide the world with a unique view of Middle Cambrian marine life. Such creatures as *Marrella splendens*, similar to a lace crab; *Aysheaia pedunculata*, a lobe-limbed "caterpillar;" and *Hallucigenia sparsa*, an odd, spinose animal that remains something of a mystery to science have become as famous in certain academic circles as the likes of Brad Pitt and Tom Cruise are to the Hollywood movie crowd. Then there's the impressively clawed, 10-centimenter-long

Tokummia kayalepsis, which ranks as the latest entry in the ever-expanding list of the shale's arthropod oddities. This peculiar bottom dwelling creature was uncovered by Canadian scientists in 2017 in a newly opened Burgess outcrop—this one in British Columbia's Kootenay National Park, located 40 kilometers southwest of Walcott's original dig.

Despite these ongoing discoveries, both academics and amateur Burgess aficionados remain as mystified as ever concerning the true roles that many of these unique animals assumed during their long-ago

KOOTENIA BURGESSENSIS (RESSER, 1942)

Middle Cambrian; Burgess Shale Formation, Campsite Cliff Member; Field, British Columbia, Canada; 4.7 cm

Although slightly metamorphosed, this specimen represents an uncommon Burgess find in complete condition. It is closely aligned to similar species that have been discovered in Utah, Idaho, and China.

lifetimes in those ancient seas. It also doesn't help clarify such matters that interested observers are often forced to examine Shale specimens exclusively through photos, or during scheduled museum visits, because Canadian authorities maintain tight control on the distribution of virtually all Burgess material—at least those fossils found after restrictions were imposed in the late-twentieth century. Yet despite the significant scientific hurdles, Burgess research continues to progress at a surprisingly sprightly pace.

"Getting hold of Burgess material has never been easy, and now it's almost impossible," said Abbott. "The funny thing is that there are supposedly something like 150,000 specimens in the Smithsonian . . . with over 50,000 of those being *Marrella*. It can make doing any kind of examination rather frustrating. But it's also part of the fun. When you do get to study a specimen from Burgess, it's magical. It's truly like dealing with a lifeform from another world."

Since the very beginning of Burgess studies, paleontologists have focused their efforts on deciphering the evolutionary riddles presented by these astounding soft-bodied arthropods, only to occasionally have their initial interpretations completely overturned by subsequent discoveries. The work of these scientists often has been dedicated to deciding within which class or order each of these tiny remnants of Earth's earliest population explosion might best be placed. Other scholarly efforts have attempted to discover which soft-tissue species may have survived long enough to provide the rootstock for future, more sophisticated animal forms.

Some academics have recently turned their attention away from the confusing and often controversial taxonomic classification of soft-bodied material and toward equally puzzling questions; why, and perhaps even more important, *how* this amazing array of faunal forms first emerged.

Why did this proliferation of seaborne activity occur after more than four *billion* years of Earth's history—the overwhelming majority of which was characterized by either nonexistent life or agonizingly slow evolutionary change—and how did it happen?

Earlier in the fossil record—perhaps best exemplified in the Precambrian by the 600-million-year-old Ediacaran layers of South Australia—no more than the slightest hint is offered concerning the variety or complexity of life to be witnessed with the onset of the Cambrian. Yet a "mere" 80 million years later, the seas are literally teeming with an incredible assortment of sophisticated creatures, including thousands of highly evolved trilobite species. Rather suddenly, for whatever still mysterious reason, just as the Cambrian began, all the oceans on Earth seemed to bloom with life. Actually, it is this blooming that signals the beginning of the Cambrian, so at least in this case it appears as if the "chicken"—or shall we say the *Sidneyia*—came before the proverbial egg. Yet it's hard not to wonder from where this astonishing assemblage of strange, soft-bodied forms first emanated. Did a wide range of biologically complex organisms exist even earlier in the Cambrian—creatures that never succumbed to the kind of quick-kill mudslides that subsequently produced the world's best-known, soft-bodied fossil fauna in the Burgess Shale?

Despite the ongoing efforts of dedicated scientific crews around the globe whose work centers on this Cambrian-age soft-tissue material, these queries remain among the great mysteries of the paleontological world. They probably will remain so until someone with enough knowledge to recognize the significance of his or her discovery stumbles on the next key to further unlocking this timeless saga. Perhaps only then will the true tale of life's invertebrate origins begin to reveal itself in all its boneless glory.

THE *PARADOXIDES* PARADOX: THE CASE FOR PLATE TECTONICS

Soon after the dawn of the Cambrian Period, thousands of distinct trilobite species already filled Earth's waters, inhabiting virtually every available maritime niche from surface to seafloor. Never, before or since, have the oceans been as evolutionarily provocative in their abundance and diversity. This was an extraordinary primal world that in many ways remains as alien to us as the moons of Jupiter.

Despite our best efforts, we may never learn exactly why the trilobite class emerged at this critical juncture in the planet's history. Almost as frustrating, we may never discover why these amazing creatures became the initial organisms on our world to feature a hard calcite exoskeleton. Yet that major morphological advance has allowed their remains to become pervasive in the fossil record. This invertebrate bounty has furnished the scientific community with a profusion of key stratigraphic markers, Paleozoic Rosetta Stones, that appear in properly aged geologic horizons around the globe. When this rich reservoir of trilobite material is compared and contrasted, especially with specimens of the same genus found in a variety of other planet-spanning locales, it can provide an unparalleled perspective on the primordial forces that helped shape our world.

Perhaps nowhere is this paleontological phenomenon more evident than when dealing with the Middle Cambrian trilobite *Paradoxides*. In fact, even a cursory examination of the various examples of that genus photographically presented within these pages will reveal something extraordinary: fossils of closely related and nearly identical members of the *Paradoxides* line have been uncovered in such disparate locations as Newfoundland, Sweden, Massachusetts, Wales, Spain, the Czech Republic, and Morocco. An obvious question then

ACADOPARADOXIDES LEVISETTII GEYER AND VINCENT, 2015

Middle Cambrian, Series 3; Agdzian Regional Stage; Jebel Wawrmast Formation, Bèche à Micmacca Member; *Morocconus notabilis* Biozone; Jebel Ougnate, near Tarhoucht, Eastern Anti-Atlas, Morocco; 23 cm

This prolific North African species is named in honor of the famed trilobite collector and researcher, Riccardo Levi-Setti.

emerges: How can fossils of such remarkably similar trilobites be found in sedimentary outcrops that are now thousands of kilometers apart? After all, it is believed that trilobites were relatively territorial animals, with their lives controlled by such factors as the depth, turbidity, and temperature of their marine environment. Few, if any, trilobite genera were equipped to traverse the open ocean, even on a limited basis. How then did the fossilized remains of essentially homogeneous trilobite species end up located in such widely separated corners of the globe? It is the *Paradoxides* paradox.

As it happens, we now know that the answer to this time-tested question is a rather basic component

of Geology 101—it all revolves around the subject of plate tectonics. Over the course of tens of millions of years, continents sitting atop massive lithospheric plates gradually slide across the Earth's outer crust like Texas T-bone steaks navigating the Teflon-coated top of a superheated frying pan. These landmasses are often propelled by seismic forces working along the key fault lines that lie directly under the planet's surface. As these continental plates continue to shift their global position ever so slowly during the passage of countless millennia—often by only 2 or 3 centimeters a year—so do the trilobite fossils contained within their sedimentary layers.

"Plate tectonics explains why the eastern 'bulge' of South America seems to fit perfectly into the western recess of Africa," said Sam Stubbs, an associate with the Houston Museum of Natural Science. "The reason is because at one time in the distant past, it did!"

Half-a-billion years ago, our world was a far different place than it is today. The now instantly familiar outlines of the continents would then have appeared virtually unrecognizable to modern eyes. At that time in Earth's history, landmasses were joined together in geographically nondescript clusters to form supercontinents, immense tracts of terra firma that carried exotic names such as Gondwana and Rodinia. These vast regions were all ostensibly aligned within the Earth's Southern Hemisphere, commonly surrounded by large bodies of water such as the Panthalassic Sea and the Iapetus Ocean.

What would one day become the heartland of North America lay along the Cambrian equator, a zone within the supercontinent known as Laurentia.

PARADOXIDES DAVIDIS (SALTER, 1863)
Middle Cambrian; St. David's Group, Davidis Zone; Porth-y-rhaw, St. David's, Pembrokeshire, Wales, UK; 23.1 cm

These trilobites are found amid treacherous seaside cliffs, which makes their collection nearly impossible. An almost identical species is found 4,000 kilometers away in Newfoundland.

PARADOXIDES SP.
Middle Cambrian; Jamtland, Sweden; 26.4 cm

One of the largest complete examples of this trilobite ever found at the Jamtland site, this impressive specimen was originally covered in a thin calcite coating that required more than 15 hours of preparation to remove.

Photo courtesy of the M. Shugar Collection

In close southern proximity was the smaller continental landmass of Baltica, which consisted of what would eventually emerge as western Europe. The sectors that would later branch off to become Wales and Newfoundland were then joined off the coast of the microcontinent of Avalonia. It was in the cool bays, coves, and basins located along this shoreline that large *Paradoxides* trilobites apparently congregated by the thousands.

Due to the turbulent volcanic activity that frequently lay behind the formation and subsequent dissolution of these continents, the world's Cambrian oceans were forced to continually expand and contract. In the process of doing so, they often flooded a significant portion of the imposing landmasses that abutted their coastlines. Indeed, during parts of the Middle Cambrian, sea levels appear to have been higher than at any previous period in Earth history, a possible consequence of the melting of large polar ice caps—which at an even earlier geologic time may have covered the entire planet like a giant snowball. Over the course of the next five-to-ten million years, this rampant and relentless oceanic activity provided trilobite genera such as *Paradoxides* with a unique opportunity to enlarge their range while evolving into many closely related species, all within a somewhat limited geographical territory.

The fossil record tells us that many of these species thrived within their chosen marine habitat. In certain places, such as the Jince fossil beds of the Czech Republic, the St. David's Group in Wales, and the Manuels River outcrops of Newfoundland, generally disarticulated *Paradoxides* carapaces are found scattered throughout the rich sedimentary layers. In the fossil stronghold of Morocco, hundreds (if not thousands) of complete *Paradoxides briareus* fossils have been commercially extracted over the past four decades.

There may well be a good reason *Paradoxides* were so pervasive in their chosen aquatic ecosystem. After all, they were among the largest of Middle Cambrian trilobites, occasionally exceeding 30 centimeters in length. Their size, along with their hydrodynamically streamlined body design, allowed them to continually control the continental shelf environs they preferred to inhabit. It is believed by many scientists that *Paradoxides* were predatory animals with a wide variety of neighboring creatures on their menu—possibly including smaller trilobites as well as a multitude of nonbiomineralized arthropods. Their dimensions, along with their apex predator status, apparently led to a disproportionate number of *Paradoxides* remains eventually becoming fossilized.

"At that time, larger trilobites had an evolutionary advantage over smaller ones," said Jason

PARADOXIDES DAVIDIS TRAPEZOPYGE (BERGSTRÖM AND LEVI-SETTI, 1978)

Middle Cambrian; Manuels River Formation; Manuels River, Newfoundland, Canada; 22.5 cm

Here is a beautiful specimen featuring an inverted free cheek (from another trilobite) perched atop its pygidium. This trilobite was once part of the famed Levi-Setti collection.

Cooper, a leading commercial trilobite explorer. "That's why large *Olenoides* trilobites can be found in significant numbers throughout the Middle Cambrian layers of the western United States, Canada, and even China, while *Paradoxides* are so

PARADOXIDES GRACILIS (BOECK, 1827)

Middle Cambrian; Jince Formation; Litavka River Valley, Czech Republic; 13.2 cm

From one of the classic Old World sites, this specimen represents a large complete example of perhaps the most renowned of all Czech trilobites.

prevalent in rocks rimming what is now the Atlantic Ocean—whether in Wales, Newfoundland, or Morocco. Apparently, in the trilobite world, sometimes bigger was better."

As members of these various *Paradoxides* species perished, their calcite-coated debris—which also featured their frequently molted external shells—was buried under millennia of mud and sand deposits. For hundreds of millions of years, the fossilizing carapaces of these large trilobites stayed trapped within their primeval resting places, first as the oceans receded and then as the continents broke apart and were slowly transported to their current locations atop the globe's outer crust. Once we recognize the incredible powers behind this tectonic process, it becomes relatively easy to comprehend why a *Pardoxides davidis* specimen uncovered along the rocky cliffs of Wales possesses a nearly identical morphological profile to a *Paradoxides davidis* now found 3,600 kilometers away amid the rugged outcrops of Newfoundland. At a special moment in Paleozoic history, some 500 million years ago, these trilobites shared an overlapping marine habitat off the coast of Avalonia.

Much the same can be said for the closely related *Paradoxides gracilis* in the Czech Republic, *Paradoxides mediterraneus* in Spain, *Paradoxides harlani* in Massachusetts, and *Paradoxides levisetti* in Morocco. During a crucial period in our planet's distant past, these species lived within the extended range of the same thriving trilobite community. Only time, combined with the unbridled forces of nature, has served to subsequently drive their fossilized remains to far-flung corners of the globe.

"The similarities between the *Paradoxides davidis* specimens found in Wales and the examples found in Newfoundland are notable," said Riccardo Levi-Setti, one of the trilobite world's most renowned figures, and after whom *Pardoxides levisetti* is named. "The rock itself may appear slightly different, but the trilobites are virtually identical. We know that these species must have lived in proximity back in the Middle Cambrian."

Throughout the prolonged course of Earth's history, clearly more than just the remnants of the *Paradoxides* line have been affected by the continent-shifting powers of plate tectonics. Even earlier in the trilobites' crawl through evolutionary time, various *Olenellus* species inhabited a relatively finite swath of the Lower Cambrian sea. Yet after millions of years of centimeter-by-centimeter movement trapped within the planet's static sedimentary layers, their distinctive fossilized forms can now be found everywhere from British Columbia, to California, to Pennsylvania, to Scotland. Similarly, species of the unusual dikelocephalid *Hungioides* have been uncovered in the Ordovician rocks of China, and other closely related examples of that rare genus have emerged from the charcoal-hued slate blocks located within Portugal's Valongo Formation. The pervasive Devonian homalonotid *Dipleura* has been discovered throughout 400-million-year-old layers in both New York and Pennsylvania, and a strikingly similar species can be found housed within the distinctive mudstone concretions of Bolivia.

Perhaps the most renowned example of this primal confluence occurs with the exotic Devonian genus *Dicranurus*. This trilobite, with its curving, ramlike cephalic "horns," ranks among the most famous Paleozoic organisms to be recovered from the fossil-filled soils of Morocco, where it is recognized as *Dicranurus monstrosus*. Yet a nearly

HYDROCEPHALUS MINOR (BOECK, 1827)
Middle Cambrian; Jince Formation, Ellipsocephalus + Rejkocephalus Zone; Jince locality, Czech Republic; each specimen: 12 cm

This attractive fossil shows a large, freshly molted carapace resting beneath its counterpart.

identical species, *Dicranurus hamatus elegantus*, is found amid similarly aged rocks in Oklahoma, now more than 8,000 kilometers away from North Africa. Apparently, some 420 million years ago, in the Middle Devonian, the land that now comprises both Morocco and Oklahoma shared an adjacent patch of ocean floor located in a shallow sea that stretched between the two early supercontinents of Gondwana and Euramerica.

Whether our focus is placed upon a Middle Devonian *Dicranurus*, or a Middle Cambrian *Paradoxides*, it seems that the primary impetus behind each trilobite's impressive worldwide distribution is the same: the dynamic intensity supplied by plate tectonics. It is this fundamental force that has played a quintessential role in shaping the geologic face of our planet: past, present, and future.

JINCE FORMATION, CZECH REPUBLIC: *A BOHEMIAN RHAPSODY*

Sometimes the work of a single scientist can change the perceptions of the entire world. Those endeavors—whether fully recognized or even acknowledged during the scientist's lifetime—often lay the foundation for a veritable revolution that occurs over subsequent years, decades, or even centuries. Charles Darwin certainly placed himself in such a position following the publication of his historic *On the Origin of Species* in 1859. So did Albert Einstein with his landmark efforts in theoretical physics during the first half of the twentieth century. Come to think of it, so did the likes of Thomas Edison, Marie Curie, Francis Crick, Isaac Newton, Nikola Tesla, Stephen Hawking, and Galileo Galilei.

And so did Joachim Barrande, admittedly, to a somewhat less dramatic extent. During the middle years of the nineteenth century, this self-taught French naturalist spent decades collecting and studying the trilobites in what was then known as Bohemia and is now the Czech Republic. At a time when activities such as geology and paleontology were often viewed as mere recreations designed to keep members of society's upper crust occupied, Barrande proved that supreme dedication to one's scientific goals was far more than an avocation—it was a noble and important life's pursuit.

Barrande (1799–1883) dedicated nearly 40 years of his life to exploring, collecting, and chronicling the roughly 320 square kilometers that house the various sedimentary formations of central Bohemia. Due to his historic efforts, the entire region is now aptly recognized in geological circles as the Barrandian Fossil Assemblage. It is an incredibly rich stratum, featuring no less than 60 prime Paleozoic outcroppings that date from the Middle Cambrian to the Middle Devonian. Within this pastoral setting dominated by rolling hills, winding rivers, and verdant farmlands, Barrande discovered, described, and subsequently named more than 4,000 fossil species. The compendium of his career work, which began emerging in 1852 as the richly illustrated treatise *Système Silurien du Centre de la Bohême*, helped revolutionize the role paleontology played in scientific studies. The wonderfully detailed trilobite drawings presented in the first volume of that 23-book series—most drawn by talented artists and lithographers functioning directly under Barrande's supervision—helped open the eyes of many to the wonders of the Paleozoic world. Few of the trilobites he showcased in that celebrated work were actually Silurian in age (his title was inspired by Sir Roderick Murchison's momentous 1839 monograph on English fossils, *The Silurian System*), but these particulars mattered little when compared to the impact Barrande's achievements had on the world's intellectual community.

For decades, those who collected or studied fossils had been treated as the proverbial red-headed stepchildren of scientific society. Their endeavors were often ridiculed, overlooked, or simply ignored by those involved with more "important" fields of academic research. But following in the wake of Murchison's groundbreaking investigations, Barrande's writings carried the study of fossils—and trilobites, in particular—to previously unimagined levels of acceptance. The depth and scope of his efforts also served as a significant influence on Darwin himself, who referenced Barrande's pioneering work in his own writings. Rather ironically, throughout the later years of his life, Barrande was a vocal proponent of Georges Cuvier's then popular "catastrophe" theory of change (which postulated that short, violent events—such as earthquakes and volcanic activity—altered the face, as well as the fauna, of the planet), a concept that ran directly contrary to Darwin's own evolutionary beliefs.

"Barrande was a very interesting individual," said Pavel Dvorak, an amateur paleontologist based in Prague who has spent many years exploring the Barradian horizons. "He was originally trained as an engineer, but when he was still a young adult he became connected to the French royal family as a tutor. When the king was forced to abdicate in 1830, he followed them around Europe, eventually ending up in Bohemia. Within a year of moving to Prague, he had launched his own fossil studies, though no one appears certain as to what drew him to that field in the first place."

Much of Barrande's work with Bohemian trilobites was conducted in a broad expanse of 510-million-year-old outcrops subsequently known as the Jince Formation. It is a geologically dynamic exposure in which a peculiar layer cake–like stratification—where bands of fossil-rich shale are interspersed between generally barren sedimentary planes—reflects the unpredictable climatic conditions that affected the region during the Middle Cambrian. Over a 20-million-year period, this area was apparently subjected to a continual series of environmental changes, transforming from a thriving deep-water marine sanctuary into a far less habitable oxygen deprived shallow-water delta . . . and then back again.

The shifting tides of mid-Cambrian seas weren't the only natural phenomenon that helped create the Jince's distinctive paleontological pedigree. In contrast to many world-renowned fossil formations—particularly those in which the fauna and flora provide clues that their surrounding Paleozoic environment had once been a warm tropical sea—the waters that covered what is now the central Czech Republic half-a-billion years ago were apparently located a significant distance south of the equator. Such a decidedly cool marine climate is also evident in other trilobite-laden strata around the world, including analogous Cambrian sites now found in Newfoundland and Wales. To the eyes and ears of leading academics, this information indicates that at a distant time in the planet's history, these now divergent sites may have once shared a similar swath of primal ocean floor.

These special environmental conditions nurtured the emergence of an incredibly diverse Middle Cambrian trilobite fauna. More than 50 species of these ancient arthropods appear in the Jince Formation, most featured in three distinct trilobite orders: Redlichiida, Ptychopariida, and Agnostida. The more common species include *Paradoxides gracilis, Ellipsocephalus hoffi, Conocoryphe cirina*, and *Hydrocephalus minor*, with the small, 2-centimeter-long *Ellipsocephalus* being found by the thousands, often on multiple specimen plates that may represent mating or molting assemblages. The Jince's rarer trilobite species include *Acadoparadoxides sirokyi, Jincella prantili*, and *Perneraspis conifrons*, with complete examples of these being almost exclusively relegated to major museum displays or to top private collections.

The region's trilobites are most often found in a heavily compacted, chocolate-colored mudstone. These distinctive outcrops have been slowly exposed over multimillennia as the meandering Berounka River cut through the tree-lined Litavka Valley near the small town of Jince. That community of 2,200 is so proud of its world-spanning scientific renown that it has adopted an illustrated and somewhat stylized rendering of a trilobite as a central component of its official city flag. When pried out of the area's fossiliferous Middle Cambrian layers and broken apart by chisel and hammer, the larger Jince Formation trilobites (which include *Paradoxides* and *Hydrocephalus* species up to 20 centimeters in length) traditionally emerge in rather flat, positive/negative "splits." The well-preserved, although often disarticulated, exoskeletons of these impressive examples display a thin calcite shell that frequently blends in color with the cocoa-tinged tone of the surrounding matrix. In contrast, many of the region's smaller genera—ranging between 3 and 7 centimeters and occasionally fossilized with a bright yellow or orange mineralized coating—have managed to retain much of their original three-dimensional convexity, despite their lengthy passage through Deep Time.

Although complete specimens have never been particularly prolific at any of the Jince sites, nearly two centuries of digging in the vicinity's outcrops have produced perhaps the most comprehensive and detailed trilobite assemblage found anywhere in the world. It is estimated that Barrande alone unearthed more than 6,000 complete and partial

ACADOPARADOXIDES SIROKYI ŠNAJDR, 1986
Middle Cambrian; Jince Formation; Czech Republic; 16.2 cm

This is one of the more unusual trilobites emerging from the fossil-rich Jince assemblage. The right genal spine is absent, indicating that this specimen is most likely a molt.

trilobites during his extensive time in the field. When paleontologists compare this amazing early Paleozoic collection to strikingly similar Middle Cambrian trilobites discovered in Eastern Canada, Sweden, Wales, and Morocco, they provide a foundational building block for one of the most important geologic theories of the last century, plate tectonics. This explains how the seven major continental plates that comprise our planet's outer shell ever-so-slowly shift their lithospheric position over time, sliding across the Earth's underlying mantle like frying eggs on a well-greased skillet.

"The trilobites of the Jince Formation are among the keystone specimens of the globe," explained Bill Barker, an American fossil dealer who has dealt extensively with Middle Cambrian

PARADOXIDES GRACILIS (BOECK, 1827)
Middle Cambrian; Jince Formation; Litavka River Valley, Czech Republic; 13.3 cm

Due to Barrande's pioneering work, this is one of the best-known and oft-studied trilobite species in the world. Even at the height of the late-twentieth-century Cold War, examples managed to appear in certain Western markets.

CONOCORYPHE SULZURI (SCHLOTHEIM, 1823)

Middle Cambrian; Jince Formation; Central Bohemia, Czech Republic; largest trilobite: 6.4 cm

Here is a fascinating group of large, blind trilobites, possibly a mating assemblage. The lack of eyes may indicate a deep-sea lifestyle.

trilobites. "Many Middle Cambrian locations feature material—especially *Paradoxides*—that are contiguous with the trilobites first found and described by Barrande while he was working in central Bohemia."

The world's awareness of Bohemian trilobites dates back more than 250 years to 1770, when a certain Professor Zeno first published a report on the assorted fossils (including a trilobite pygidium) he had discovered while exploring throughout the Prague region. Various subsequent scientific manuscripts made their appearance during the nineteenth century, with these either directly or indirectly referencing the area's vast paleontological reserves. These later efforts were, of course, highlighted by Barrande's own multivolume work, which is rumored to have personally cost him nearly $15,000 (the equivalent of more than $400,000 today) to fashion, publish, and distribute throughout Europe. He offset a great deal of this exorbitant expense by selling many of his best study specimens—some being the figured trilobites presented on the pages of *System Silurien*—with most of these examples eventually forming the core trilobite collection housed in Prague's National Museum.

Other original Barrande trilobites quickly made their way to institutions around the globe. These trilobite-centric transactions were headlined by a significant mid-nineteenth-century acquisition by the famed Louis Agassiz for Harvard's Museum of Comparative Zoology, an agglomeration that still represents the most comprehensive collection of Bohemian trilobites housed outside of Europe. Such well-publicized commercial ventures—as well as their resulting museum displays—soon triggered a worldwide fascination with these ancient arthropods. In fact, that interest escalated to a point where only a few years after Barrande had finished conducting his pioneering research, quarrymen working in many of the same fossil-rich Central European formations were renowned (if not reviled) for their willingness to find, piece together, and then sell what were essentially disassociated trilobite parts. In the process of doing so, these financially motivated operatives often ended up creating chimera-like monstrosities unknown to both science and the Cambrian seas.

Despite the planet-spanning attention drawn to the area's Paleozoic horizons due to Barrande's historic efforts—as well as to some of the locality's more notably nefarious paleontological practices—by the middle of the twentieth century, following the advent of World War II, the chance to experience hands-on at the source contact with any of the region's more exotic trilobite species came to an abrupt halt. That opportunity would only begin again some five decades later following the fall of the Soviet Union in the late-1980s.

In the interim, the inherently secretive nature of the postwar Soviet satellite state of Czechoslovakia strictly limited exportation of the nation's natural resources, including fossils. At certain strange moments, the government seemed to impose a ban on scientific thought escaping national boundaries, let alone the actual specimens that may have helped nurture those academic concepts and philosophies. During the Cold War "dark ages," a few representative trilobites—mostly semicomplete *Paradoxides* molts missing their free cheeks—managed to filter out of the region and reach Western shores. At the same time, copies of Barrande's original manuscripts could still be found in many museum reference rooms, and photographs of unusual Bohemian trilobite species remained available through the auspices of leading European collectors and institutions such as Harvard's Museum of Comparative Zoology.

For nearly half a century direct access to the nation's rich abundance of trilobites was strictly controlled by government edict. It was rare indeed for a Western scientist to be afforded the opportunity to wander through the extensive Barrandian collections held within the confines of Prague's National Museum. However, as soon as the walls of communism came tumbling down, both literally and figuratively, and the doors of the Czech Republic were thrown open to outsiders, the incredible beauty, diversity, and significance of these Bohemian trilobites began once again to be fully appreciated by collectors and academics around the globe.

In recent years, new laws have been implemented by the Czech government, and some are directly designed to restrict digging for fossils in the Jince Formation—now viewed as a National Historic Site. At the same time, greater access to museum collections and additional scientific fieldwork and research has greatly increased the world's cognizance of the entire Barrandian Lagerstätte. Informative internet sites, richly illustrated books, full-color brochures, and comprehensive DVDs have all now appeared on the global market, each highlighting some aspect of the inherent elegance or the scientific importance of the nation's rich Middle Cambrian trilobite reserves.

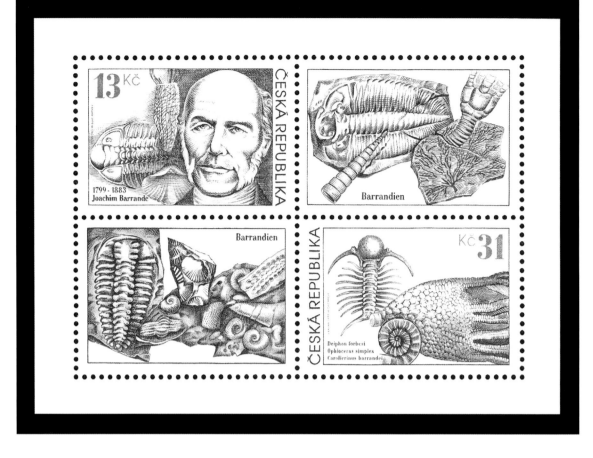

In 1999, the Czech Republic honored the work of Joachim Barrande through a series of trilobite-centric postage stamps.

"There has definitely been a reawakening of interest in this material," Dvorak said. "Many in today's generation don't remember the time when it was difficult for even Czech scientists to study our own trilobite resources. These days it is encouraged—not only for us, but for scientists from around the world, as well."

Now, nearly two centuries after Barrande's historic volumes first made the entire academic community aware of the incredible faunal diversity contained within Bohemia's fossil outcrops, there is an ongoing renaissance surrounding both his writings and the trilobites that inspired them. In 1999—the year that celebrated the two-hundredth anniversary of his birth—a limited edition, hardcover reprint of *Système Silurien du Centre de la Bohême, Vol. 1* was issued by the National Museum . . . and almost immediately sold out. That same year, the Czech Republic released a highly popular block of four commemorative stamps, one featuring Barrande's portrait and the other three presenting some of his more unusual trilobite discoveries.

Many of the species names originally attributed by Barrande have since been changed, but the details presented in his writings, as well as the brilliance of the trilobite illustrations featured in *Système Silurien*, remain unchallenged. Clearly Joachim Barrande stands on a lofty plateau, one reserved for only the most visionary of academic explorers. His studies may never have attained the Earth-changing heights of a Darwin or an Einstein, but without his pioneering efforts the science of paleontology, as well as the theory of plate tectonics, would perhaps never have garnered the degree of credibility and renown they currently enjoy.

"Every time I hold a trilobite from the Czech Republic, whether it is a small *Ellipsocephalus* or a large *Paradoxides*, it's hard not to think of Barrande," said Dvorak. "How fortunate he was to be the man who first found and studied such amazing specimens, and how fortunate we are to have his work to learn from and appreciate."

RAPID REPORTS

Parker Quarry, Vermont

Located in the outskirts of a small town in northern Vermont is a legendary trilobite-bearing formation known as the Parker Quarry. Renowned American paleontologists, including James Hall and Charles Walcott, examined specimens emanating from the locale's 520-million-year-old Lower Cambrian layers as far back as the late nineteenth century, and a magnificent collection of the site's *Olenellus* trilobites (some up to 15 centimeters in length) has been housed in the Smithsonian ever since. Late in the twentieth century it was believed that the Parker Quarry had either been played out by overzealous digging in previous decades or had simply been lost to the whims of modern development. In early 2012, however, a small team of determined trilobite enthusiasts studied old reports regarding the

site's possible location and commenced a detailed investigation designed to find out if the layers still existed and if they could once again be explored. It took more than two years of visits to rural Vermont before an outcrop of what was believed to be the original Parker Quarry was discovered, located squarely in the middle of an active cow pasture.

Sirius Passet, Greenland

Of all the distant, isolated and hard to reach destinations described in these pages, perhaps none can rival the sheer seclusion presented by the Sirius Passet Lagerstatte, a Lower Cambrian outcrop that borders the Arctic Sea along the northernmost coast of Greenland. Discovered in 1984 by explorers working in conjunction with the Geological Survey of Greenland—and named in honor of that area's renowned Sirius Dog Sled Patrol—so far six small outcrops of the 518-million-year-old Buen Formation have been revealed among the sedimentary exposures that run adjacent to the ice-filled J. P. Koch Fjord. In addition to a variety of sponges, brachiopods, and soft-bodied arthropods, the fauna presents a limited selection of trilobite species, including *Buenellus higginsi*, representing some of the earliest examples of the trilobite line. Preservation at these sites has so far proved to be somewhat problematic—with the fossils generally appearing as poorly preserved elements of large microbial mats—yet the Sirius Passet locale still holds the promise of revealing some startling discoveries, especially as additional fieldwork is conducted in this remote corner of the globe.

Kinzers Formation, Pennsylvania

The Kinzers Formation in Lancaster County, Pennsylvania, has long been renowned for its distinctive Lower Cambrian trilobite fauna, much of which hails from a series of closely aligned quarries that have been explored and collected since the late nineteenth century. Of specific interest to collectors are the beautifully preserved examples of *Olenellus getzi* that emanate from a sedimentary exposure that appears near the local hamlet of Rohrestown. These specimens, which can reach 17 centimeters in length and are often preserved in a bright yellow or orange limonite, occasionally display soft-tissue preservation (mostly antennae) as well as wonderfully detailed dorsal morphology. Another Kinzers species near and dear to the hearts of trilobite enthusiasts is *Wanneria walcottana*, named in honor of the seemingly ubiquitous Charles Walcott. Although complete *Wanneria* specimens are now rarely if ever found (the area's Getz and Brubaker quarries have been closed since the 1970s, with the latter now covered by a parking lot), in the later years of the twentieth century, these large, elegant fossil arthropods appeared in significant numbers, becoming staples of museum and private collections throughout the Northeast.

Pioche Shale Formation, Nevada

Just a quick hop from the notorious UFO haven of Area 51, and a mere three-hour drive from the glittering lights of Las Vegas, there has long been a bit of added Paleozoic panache surrounding the Lower Cambrian outcrops that comprise Nevada's Pioche Shale Formation. The environs in this part of Lincoln County present more than their share of challenges to those in search of fossils, especially with summer temperatures that routinely reach 115° Fahrenheit (46° Celsius), but the results derived from exploring this fascinating locale are often worth the inherent difficulties. Captured within these yellow-tinged, 514-million-year-old sedimentary layers are key elements detailing one of early Earth's most important transitional stages, as the Lower Cambrian fauna slowly began

to fade and eventually were replaced by a flurry of new Middle Cambrian genera. Featuring an exotic array of closely related trilobite species, including *Bristolia bristolensis*, *Bristolia nevadensis*, *Olenellus fowleri*, and *Olenellus chiefensis*, all of which are preserved here in a golden-hued hematite, the trilobites of the Pioche Shale represent some of the most dramatic and aesthetic fossils to be found anywhere in the world.

Manuels River, Newfoundland, Canada

Trilobites were initially discovered near Conception Bay, Newfoundland, in 1874 by a survey team working under the auspices of the Geological Survey of Canada. But by the time Riccardo Levi-Setti first visited this then long-abandoned Middle Cambrian outcrop in the mid-1970s, the rocky expanse along the Manuels River had degenerated into a makeshift garbage dump. After removing the rusting hulks of refrigerators and washing machines, Levi-Setti was able to uncover 510-million-year-old mudstone layers that were filled with magnificent examples of large *Paradoxides davidis*, along with a smattering of other related trilobite species, such as *Anopolenus henrici*. Not only were these specimens aesthetically pleasing (as showcased in Levi-Setti's subsequent trilobite books, which frequently highlighted this Manuels River material), but they also proved to be of key scientific importance. When compared to remarkably similar *Paradoxides davidis* specimens from Wales (now some 3,200 kilometers away), these trilobites provided additional support to the concept of plate tectonics, which explains the movement of continents over time. Though trilobite collecting is now banned at the location, a small museum that features a variety of Levi-Setti's discoveries has opened near the Manuels River site.

McKay Group, British Columbia, Canada

Whenever the trilobites of British Columbia are considered, images of the historic Burgess Shale and its world-renowned trove of Middle Cambrian material naturally, and rightfully, spring to mind. That high-altitude outcrop of the Stephen Formation stands as perhaps the most studied and lauded invertebrate fauna in paleontological history. However, this rugged, mountain-strewn province along Canada's Pacific Coast should also be recognized for its variety of intriguing Lower and Upper Cambrian locations, many of which sporadically appear throughout the entire region. Magnificent *Elliptocephala* and *Wanneria* specimens emerge from the area's oldest fossil-bearing strata—the Rosella Formation, found within the remote Cassiar Mountains, housing the former and the Eager Formation in Cranbrook being renowned for the latter. The Upper Cambrian is also well represented in British Columbia, particularly by the diverse and prolific McKay Group. In this series of limestone outcrops, dozens of trilobite species, including *Orygmaspis contracta* and *Labiostria westropi*, are often found in calcified nodules that produce beautifully detailed positive/negative splits.

Weeks Formation, Utah

Due to both the unique preservation and stunning diversity exhibited by its more than two dozen species of described trilobites, Utah's 500-million-year-old Weeks Formation represents one of the most captivating fossil repositories in the western United States. Featuring a distinct fauna of generally diminutive trilo-types that rarely exceed 4 centimeters in length—including *Tricrepicephalus texanus*, *Meniscopsia beebei*, and *Norwoodia boninoi*—the Weeks was long thought to represent

OLENELLUS SCHUCHERTI (RESSER AND HOWELL, 1938)

Lower Cambrian, Series 2, Dyeran; Eager Formation; Cranbrook, British Columbia, Canada; 5.2 cm

This primitive genus is notable for its worldwide distribution. Various *Olenellus* species have been found in Nevada, Pennsylvania, Newfoundland, and Scotland, as well as at this site in western Canada.

a well-defined Upper Cambrian biozone. However, recent studies—along with the recovery of a rare species of large *Olenoides*—indicate that it is, in fact, one of the last bastions of the Middle Cambrian to be found anywhere in North America. The trilobites discovered here are often preserved in a

(OVERLEAF) **WANNERIA SP.**

Lower Cambrian; Eager Formation; Cranbrook, British Columbia, Canada; 10.7 cm

During the later years of the twentieth century, a layer featuring these impressive trilobites was uncovered in the Canadian Rockies. They are still undergoing scientific study and classification.

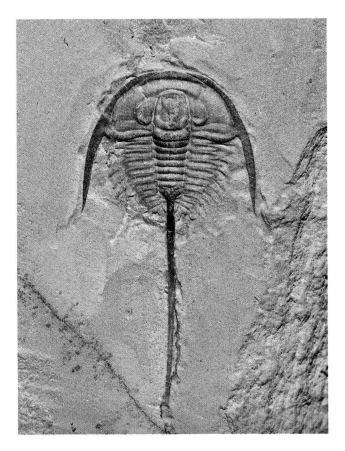

thick black or brown calcite embedded in a matrix that can vary from tan to pink in color, factors which make Weeks material (which also includes the large aglaspid, *Beckwithia typa*, along with an assortment of generally small, soft-bodied organisms) among the most collectible in the world. In addition, following a decade during which the formation's shale layers were commercially mined for use as garden paving stones, recent legislation has placed much of the area's Paleozoic outcroppings on a "restricted" list, serving to further increase the lure of any and all trilobites that hail from the Weeks.

CEDARINA SCHACHTI (ADRAIN, PETERS, AND WESTROP, 2009)

Middle Cambrian; Weeks Formation; Millard County, Utah, United States; 6.1 cm

Trilobites from the Weeks Formation were particularly coveted during the final years of the twentieth century when digging at the site became limited.

CAMBRIAN TRILOBITE
PHOTO GALLERY

(LEFT) OLENOIDES SP.

Middle Cambrian; Marjum Formation; Utah, United States; 8.4 cm

Closely aligned with *Olenoides pugio*, this specimen displays slightly different morphological features.

(BELOW) MODOCIA SP.

Middle Cambrian; Wheeler Formation; Utah, United States; 6 cm

Even in oft-studied formations, unexpected and undescribed species such as this one may still emerge.

(OPPOSITE PAGE) ELYX AMERICANUS HOWELL, 1932

Lowest Middle Cambrian; St. Albans Formation, *Centropleura vermontensis* beds; Franklin County, Vermont, United States; 4 cm

This specimen was found amid loose debris in 2012 while explorers were searching for the famed Parker Quarry.

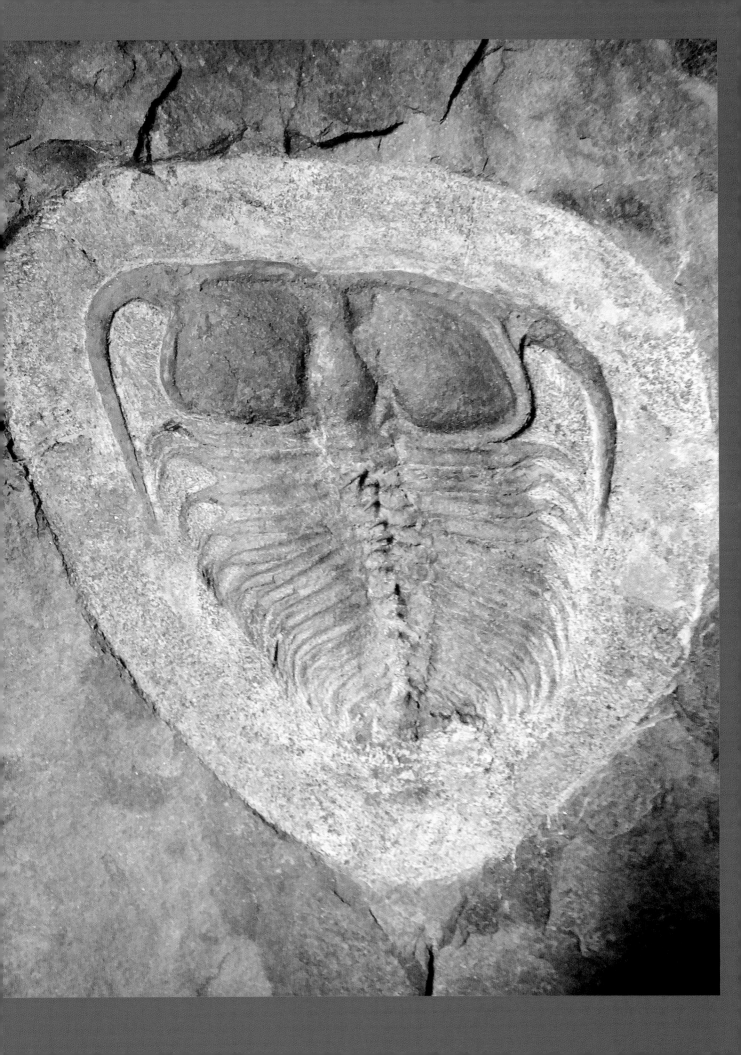

(ABOVE) *IRVINGELLA MAJOR* ULRICH AND RESSER, IN WALCOTT, 1924

Upper Cambrian, Furongian; McKay Group, Elvinia Zone; Kootenay Mountains, British Columbia, Canada; 8.5 cm

This represents one of the more dramatic trilobites emerging from a commercial dig that has been conducted over the last decade high up in the Kootenay Mountains.

(OPPOSITE PAGE) *OLENOIDES VALI* ROBISON AND BABCOCK, 2011

Middle Cambrian; Upper Wheeler Formation; Drum Mountains, Utah, United States; 10.1 cm

This species was described in a 2014 scientific paper that focused on the diversity of Utah's fossiliferous material. It was named in honor of Val Gunther, whose family has pioneered trilobite field work for more than 60 years.

(RIGHT) *NEVADELLA N. SP.*

Lower Cambrian; Between Laudonia and Elliptocephala Zones; Poleta Formation; Montezuma Range, Nevada, United States; 4.1 cm

An unusual and as yet unclassified species hailing from a remote high-altitude corner of Nevada. This specimen represents one of the earliest-known trilobite species in North America.

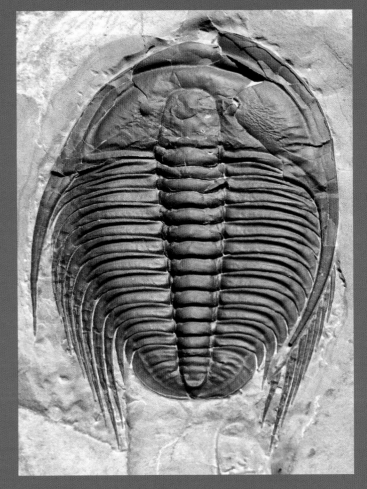

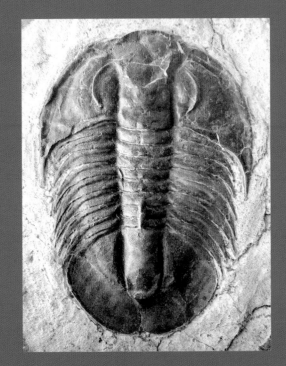

(ABOVE, LEFT) *OLENOIDES SKABELUNDI* **ROBISON AND BABCOCK, 2011**

Middle Cambrian; Weeks Formation; House Range, Utah, United States; 16.5 cm

This impressive trilobite helps prove that the Weeks Formation is a Middle Cambrian outcrop rather than—as had long been assumed—a representative of the Upper Cambrian.

(ABOVE, RIGHT) *ORYGMASPIS (PARABOLINOIDES) MCKELLARI* **CHATTERTON AND GIBB, 2016**

Upper Cambrian, Furongian; McKay Group, Taenicephalus Zone; Cranbrook, British Columbia, Canada; 6 cm

Over the last two decades, this high-altitude outcrop has produced hundreds of complete trilobites, featuring dozens of unusual species. This is one of the more visually attractive specimens to emerge from that dig.

(BOTTOM, RIGHT) *GLOSSOPLEURA YATESI* **(ROBISON AND BABCOCK, 2011)**

Middle Cambrian; Langston Formation, Spence Shale Member; Utah, United States; 4.2 cm

Note the small axial spines that appear on this trilobite's carapace.

(OPPOSITE PAGE) *NEVADELLA ADDYENSIS* **(OKULITCH, 1951)**

Lower Cambrian, Dyeran; Poleta Formation; Montezuma Range, Esmeralda County, Nevada, United States; 5.2 cm

Even very early in their crawl through Deep Time, trilobites had already achieved a degree of evolutionary perfection with their streamlined design and hard outer shells.

(ABOVE, LEFT) *OLENOIDES PARAPTUS* ZHAO, AHLBERG AND YUAN, 1994

Lower Middle Cambrian, Taijiangian Stage; Kaili Formation, Oryctocephalus indicus Zone (Kaili Biota); Taijiang County, Guizhou Province, China; 6.3 cm

Olenoides represent one of the keystone genera of the Middle Cambrian. Species have been found in such disparate locations as Utah, British Columbia, Siberia, China, and in the U.S. state of Georgia.

(ABOVE, RIGHT) *ELLIPTOCEPHALA SP.*

Lower Cambrian; Rosella Formation, Atan Group; British Columbia, Canada; 16.6 cm

This is a beautiful large example from a remote locale; both the species and the location are in need of further academic investigation.

(BOTTOM, LEFT) *LOCHMANOLENELLUS TRAPEZOIDALIS* WEBSTER AND BOHACH, 2014

Lower Cambrian (Series 2), lower Dyeran Regional Stage; Middle Member of the Poleta Formation; Montezuma Mountains, Esmeralda County, Nevada, United States; 14.4 cm

The pronounced pairing of genal spines atop this creature's head account for its highly unusual appearance. This species was part of a major scientific study—and subsequent taxonomic revision—in 2014.

(OPPOSITE PAGE) *TRICREPICEPHALUS TEXANUS* (SHUMARD, 1861)

Middle Cambrian; House Range; Weeks Formation; Millard County, Utah, United States; largest trilobite: 5.2 cm

This highly unusual "triple" ranks among the more eye-catching examples of this species to emerge from the Weeks Formation exposures.

(ABOVE, LEFT) *MYOPSOLENITES BOUTIOUITI* GEYER AND LANDING, 2004

Middle Cambrian, Tissafinian Stage; Jebel Wawrmast Formation; Brèche à Micmacca Member; Jebel Ougnate, Tarhoucht, Morocco; 12.2 cm

A large, elegant trilobite drawn from quarries located near the famed *Paradoxides* layers of central Morocco; complete examples of this species have only been seen commercially since 2010.

(ABOVE, RIGHT) *AMECEPHALUS ALTHEA* (WALCOTT, 1916)

Middle Cambrian; Langston Formation, Spence Shale Member; Box Elder County, Utah, United States; 5.1 cm

This is a long-known but only recently described species that was the subject of a scientific revision in 2015.

(OPPOSITE PAGE) *PEACHELLA IDDINGSI* (WALCOTT, 1884)

Lower Cambrian; Carrara Formation, Echo Shale Member; Tecopa Hot Springs, California, United States; 3 cm

With its strangely inflated genal spines, this 520-million-year-old species ranks among the most unusual trilobites in the world.

(ABOVE) ELLIPTOCEPHALA SP.

Lower Cambrian, Series 2, lower Dyeran; Poleta Formation; Montezuma Mountains, Nevada, United States; 7.5 cm

This genus is currently undergoing scientific study alongside several distinct species ranging in similarly aged strata from Nevada to British Columbia.

(OPPOSITE PAGE) **OLENOIDES SUPERBUS (WALCOTT, 1908)**

Middle Cambrian; Marjum Formation, House Range; Millard County, Utah, United States; 12.4 cm

This is a dramatically prepared specimen of an unusual Middle Cambrian trilobite. A pocket of these once rare bugs was uncovered in 2010, with more than 20 complete specimens subsequently emerging.

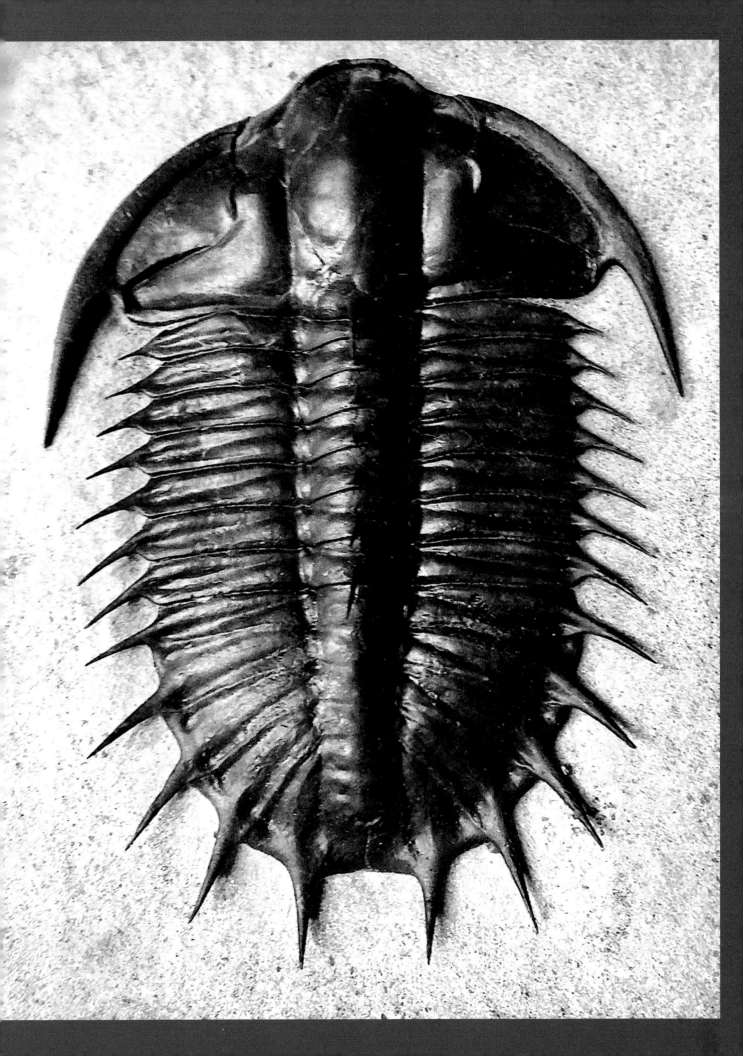

(ABOVE) *HAMATOLENUS (HAMATOLENUS) VINCENTI* GEYER, 2004

Middle Cambrian, Series 2, Age 4, Tissafinian Regional Stage; Jbel Wawrmast Formation, Brèche à Micmacca Member; Morocconus notabilis Zone; Morocco; 4.9 cm

Morocco has become renowned for its abundant Devonian fauna, but it also possesses a notable variety of Cambrian trilobites. Note the elongated thoracic segment on this specimen.

(OPPOSITE PAGE, TOP) *HOUSIA CANADENSIS* (WALCOTT, 1912)

Upper Cambrian, Furongian; McKay Group, Elvinia Zone, *Wujiajiania lyndasmithae* to *Wujiajiania sutherlandi* subzones; Cranbrook, British Columbia, Canada; 4.7 cm

This is a particularly well-preserved example of an unusual species. The formation from which it emanates has been the focal point of a major commercial operation since 2002.

(OPPOSITE PAGE, BOTTOM LEFT) *YINITES YUNNANENSIS* ZHANG W., 1966

Lower Cambrian, Series 2, Stage 3 (Qiongzhusian Regional Stage); Hongjingshao Formation; Southeastern Yunnan Province, China; 5.4 cm

Strange, early trilobites such as this were known for decades from Chinese literature, but only in recent years have complete examples been seen by the outside world.

(OPPOSITE PAGE, BOTTOM RIGHT) *EMIGRANTIA N. SP.*

Lower Cambrian, Dyeran, middle to upper Bristolia Zonule; Carrara Formation, Thimble Limestone Member; Nopah Range, Inyo County, California, United States; 3.2 cm

Here is an uncommon member of the Biceratopsidae family, representing one of the more bizarre lines on the early trilobite evolutionary tree. There has been some damage to the tips of this specimen's spines.

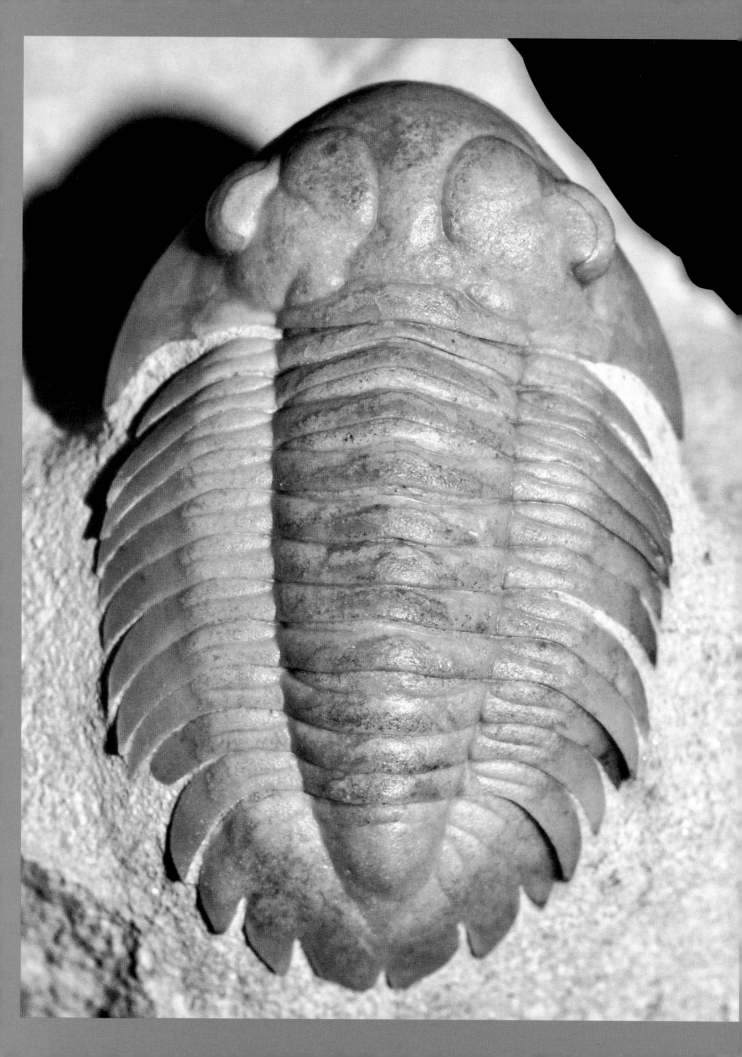

THE ORDOVICIAN PERIOD 2

485–444 Million Years Ago

ASERY HORIZON, VOLKHOV RIVER, RUSSIA:
FROM RUSSIA WITH BUGS

JUST ABOUT EVERY SATURDAY morning from early March until late October (and more frequently when Russia's harsh winters allow) Stanislav Pogorelskyi and his two college friends get up before the crack of dawn, pull on their warmest sweaters and heavy weather-resistant boots, and head into the Volkhov river valley on the outskirts of St. Petersburg. It takes the three geology graduate students a little more than 90 minutes by public transportation to cover the 100-kilometer distance required to reach the still mostly pristine environs of the valley in which they plan on spending the next 10 hours. Their long day includes a healthy 8-kilometer hike over the area's sparse and craggy landscape and a quick nutritious lunch that Stan's mother prepared the night before. But their primary mission in the valley isn't hiking or picnicking—it's collecting fossils, Ordovician-age trilobite fossils, to be exact.

With tireless enthusiasm, the trio will move rapidly but carefully through the hillsides that encircle these 460-million-year-old sedimentary formations known collectively as the Asery Horizon. Occasionally

PLATYLICHAS SP.
Lowest Upper Ordovician, Caradocian; Kukruse Regional Stage; St. Petersburg, Russia; 2.5 cm

This genus was known strictly from scattered parts for more than a century. Recent discoveries, however, have provided both science and collectors with occasional complete examples.

they will stop to jab at a rock outcrop with their sturdy geologic hammers, hoping to see if one of the area's dozens of trilobite species has been partially exposed by the twin forces of weathering and erosion. Often they will spot little more than a small speck of caramel-colored shell jutting out of the surrounding strata. It won't be until they begin freeing that speck from its covering matrix that they will know if their discovery is just another distracting fragment or what they've been searching for—a complete trilobite specimen.

Carefully, they will chip away the rock encompassing any promising find, intending to keep as much of the trilobite in one piece as possible. But the hard, chalky limestone rarely breaks cleanly. Fossils usually are removed in anywhere from 2 to 10 pieces, only to be painstakingly reassembled later. They meticulously wrap their best discoveries in protective layers of newspaper before placing them in their backpacks. There they will safely stay alongside perhaps a half-dozen other "keepers" found during the long day's sojourn and will be carried back to the group's St. Petersburg laboratory.

Their lab is little more than a basement room that the industrious trio have converted into a makeshift trilobite research and preparation station. In this small, brightly lit facility, under the gaze of a pair of high-powered binocular microscopes, the delicate pieces of these ancient sea creatures (generally between 4 and 15 centimeters in length) will be diligently extracted from their rock encasements, reassembled with both skill and imagination, and returned to much of their original Ordovician grandeur.

It requires at least five hours of exacting labor, with tools ranging from dental picks to air-powered pneumatic drills, to properly remove the hard limestone matrix that too often adheres to even the smoothest trilobite surface. Particularly exotic examples, such as those featuring rows of fragile, freestanding spines, can take considerably longer to prepare and restore. But time and patience are two commodities that those who seek out these Paleozoic treasures seem to possess in abundance.

"The work of finding and then preparing the trilobites is very difficult, but we enjoy it," stated Pogorelskyi. "We are all students of geology and fossils, so we learn a great deal as we dig in different places within the valley and find more material. Each year, new layers are exposed through natural weathering, and new opportunities for discovery arise. Our goal is to find a new species."

For decades prior to the fall of the Soviet Union in the late 1980s, rumors of stupendous Russian fossil reserves had been reported by paleontologists lucky enough to have visited behind what was then known as the Iron Curtain. Tightly sequestered backrooms in major Soviet museums were said to be bursting at the seams with material collected everywhere from the flat tundra of eastern Siberia to the hilly cliffs of Estonia. Western scientists who had been invited on rare Russian field expeditions between 1917 and 1989 often returned with glowing reports of strata bulging with ammonites, trilobites, and vertebrate material of all orders and ages. However, most Soviet scientists of the era were strongly advised to keep their interests focused on more pressing matters than the collection and study of fossils. And with the Russian government then carefully regulating the export of their nation's natural resources, few of these Eastern European riches ever found their way onto the world market.

ESTONIOPS EXILIS (EICHWALD, 1857)

Upper Ordovician, lower Caradocian; Kukruse Regional Stage; Viivikonna Formation, Alekseevka Quarry; St. Petersburg region, Russia; 3.6 cm

Unusually detailed preservation distinguishes this small but pristine specimen. Russia's Upper Ordovician quarries produce a trilobite fauna distinctly different from their better-known Middle Ordovician counterparts.

By the early 1990s, however, all of that had begun to change. At major fossil and mineral shows such as those held annually in Tucson, Tokyo, and Munich, magnificent and unique Russian specimens ranging from huge split and polished ammonite halves, to complete 3-meter-tall cave bear skeletons, to large chunks of Siberian mammoth hide (the long-deceased beast's dark red hair still attached) began to surface. But of all the Russian fossil material that suddenly started appearing for sale in both Western and Asian markets, perhaps none drew as much attention from both the academic and collecting communities as the fantastically freakish assortment of Ordovician trilobites that were being pulled out of those Asery Horizon hillsides along the Volkhov River.

With their shiny, three-dimensional, toffee-hued carapaces contrasting attractively against a light tan matrix, these ancient arthropods soon became paleontological sensations. They were featured in major museum displays around the globe and were marketed as part of successful natural history auctions held by the likes of Sotheby's and Christie's. One particular species, the strange yet relatively prevalent *Neoasaphus kowalewski*—with eyes sitting atop spindly stalks that frequently reached a length of 4 centimeters or more—quickly became a "must have" for every trilobite collector from St. Petersburg to St. Louis. That alien-looking species soon revealed itself to be just the forerunner of what would rapidly prove to be nothing less than a trilobite-themed Russian revolution. Indeed, during the three-plus decades since the fossils of the Asery Horizon first began appearing in the

PARACERAURUS ACULEATUS (EICHWALD, 1857)
Upper Ordovician, Caradocian; Kukruse Regional Stage; Viivikonna Formation, Alekseevka Quarry; St. Petersburg region, Russia; 10.4 cm

This beautifully prepared specimen is one of the rarer trilobites emerging from the St. Petersburg quarries.

METOPOLICHAS HUEBNERI (EICHWALD, 1843)
Middle Ordovician, Lower Llandeilian; Lasnamägi Regional Stage; Porogi Formation, Vilpovitsky Quarry; St. Petersburg region, Russia; 12.1 cm

This is one of the largest specimens yet found of this distinctive variety of lichid.

international spotlight, an escalating number of trilobite genera have emerged from the bountiful Volkhov River region. Each year has produced a series of previously unseen species from adjacent Ordovician quarries—with every new trilobite discovery seemingly determined to outdo its Paleozoic predecessor in terms of its morphological eccentricities. Many of these specimens have subsequently proven themselves to be among the most bizarre, beautiful, and coveted trilobites found anywhere in the world.

"Russian trilobites changed everyone's perspectives on what a specimen could and should look

PSEUDOSPHAEREXOCHUS ORVIKUI RALF
MÄNNIL, 1958

**Upper Ordovician, Caradocian; Kukruse Regional Stage;
Viivikonna Formation; St. Petersburg, Russia; 3 cm**

The "ivory" tone of this trilobite and the brick-red matrix color
are indicative of new material emanating from the Viivikonna
Formation.

Ordovician trilobites, many of which share similar genera but subtly divergent species. In recent
years, scores of complete specimens have emerged
from these layers—many of trilo-types previously
known only from partial, broken, or disarticulated
finds. Whether they were spinose odontopleurids,
smooth illaenids, or sleek harpids, each represented a magnificent (and often expensive) example of the diverse fauna that once inhabited this
area's shallow Paleozoic seas.

Despite their high cost of acquisition (usually
priced between $200 and $800 on the retail market, but their commercial value has occasionally
soared into four- and even five-figure territory),
these alluring trilobites have quickly managed to
enchant virtually everyone with whom they've
come in contact—whether that person is a university-trained academic or a Wall Street broker
interested in natural history.

If that Middle Ordovician paleontological
bounty wasn't enough to satisfy the creature cravings of any true-blue trilobite enthusiast, in the
second decade of the twenty-first century, a small
number of freshly opened Upper and Lower Ordovician quarries began drawing a profusion of long
unseen trilobite species from the surrounding
Russian hillsides. These discoveries—including
such once rare bugs as *Paraceraurus aculeatus,
Cybelella coronata*, and *Reraspis plautini*—have
served to further expand the species count attrib-
utable to the St. Petersburg region's fossiliferous
formations, which had reached an impressive tally
of well over 100 by 2021. In fact, it has been hinted
that the often fragmentary remains of at least one
new trilobite species is being uncovered annually as more and more diggers continue to search
through these nearly half-billion-year-old layers in
their quest for Paleozoic "gold."

Whether they are Upper, Middle, or Lower
Ordovician, the allure of these striking Russian
trilobites is immediate and obvious. There is

like," said Martin Shugar, a major New Jersey–
based trilobite collector. "They are almost all stunning to look at, and because of their size, color, and
appearance, they've become favorites of collectors
everywhere."

The fossil-rich Asery Horizon presents a series
of distinctly different, yet stratigraphically connected quarries including the Vilpovitsky, Putilovo,
and Gostilitsy, which together span approximately
two million years of geologic time. It is not surprising that each of these fossil-filled repositories
features its own unique assemblage of Middle

CYBELELLA CORONATA (SCHMIDT, 1881)

Upper Ordovician; Kukruse Stage; Alekseevka Quarry; St. Petersburg, Russia; 4 cm

During the initial decades of the twenty-first century, the Alekseevka Quarry has yielded some of the most dramatic trilobites ever seen.

Nieshkowskia timuda, a cheirurid with a strange, hooklike projection emanating from its thorax; *Megistaspidella triangularis*, an asaphid with a triangular head and a long, spiked rostrum; *Cybele bellatula*, an encrinurid with 3-centimeter-long eye stalks as thin as angel hair pasta; and *Hoplolichas furcifur*, a lichid species with a bulbous, quill-covered cephalon. These trilobites, among many others, have continued to amaze both amateur collectors and professional paleontologists around the globe with their striking variety of spines, shapes, and sizes.

Getting those trilobites to appear as pristine as they so often do is far from an easy task. Indeed, if tales emanating from various

ASAPHUS KOWALEWSKII LAWROW, 1856

Middle Ordovician, Middle Llanvirnian; Middle Asery Regional Stage; Duboviki Formation; Volkhov River, St. Petersburg region, Russia; 7.5 cm

This bizarre Russian species features eyes perched atop 4- to 6-centimeter-long stalks. Relatively common, it has become the "must have" Russian species for collectors around the world.

St. Petersburg–based trilobite laboratories are to be believed, it often takes preparators upward of 40 hours to fully expose the pincushion-like proliferation of barbs attached to each *Hoplolichas* specimen. Even properly preparing the most basic *Asaphus* species can require a full day's focused labor, depending, of course, on the degree of energy required to place various bits of "wandering" trilobite shell back where they rightfully belong. But the results are worth the toil, and increasingly effective prep equipment has allowed these Russian craftspeople to create extraordinary presentations, many of which now stand as true works of Paleozoic art.

Unfortunately, in lockstep with the progressively more dramatic appearance of these trilobites—with some now being introduced in a totally freestanding state, anchored to their matrix base only by a thin pedestal of reinforced metal—have come a series of corresponding questions and concerns regarding the true authenticity of these highly appealing, 460-million-year-old maritime remnants. Increased worldwide interest in the Asery Horizon's Ordovician opulence has ignited a corresponding surge in what might kindly be termed "unsavory" business practices by some operating along the Russian trilobite pipeline.

In the early years of the twenty-first century, reports of rare St. Petersburg area trilobite species being assembled from various similarly sized but independently found parts began to dominate conversations among serious trilobite enthusiasts. Rumors of particularly intriguing (and valuable) species such as *Boedaspis ensifer*—a large, spiny variety of odontopleurid, prime examples of which have sold for $10,000—being molded in calcite-infused plastic and then artfully placed on slabs of actual Volkhov matrix began to further alarm collectors around the world. Occasionally, overzealous and unenlightened Russian lab workers have constructed their various piecemeal trilobite concoctions into genus-bending monstrosities that would never have seen the light of day in the Ordovician seas. More than a few of these flagrantly fabricated specimens made it into prominent private collections and major museum displays before whistleblowers began to vociferously complain about such science defying schemes.

As might be expected, leading Russian trilobite dealers either quickly denied these charges of fakery or merely laughed at the notion that such unscrupulous activities could be taking place right under their supposedly knowledgeable noses. Their responses to such "outrageous" claims were even occasionally masked behind incongruous statements about merely trying to provide their ever-more demanding legion of international customers with "what they want" in terms of exotic new species. Even after these duplicitous practices were fully exposed in the early 2010s, this cavalier attitude raged on, but in a somewhat more sophisticated manner. In more recent times, these dealers continue to occasionally "enhance" both manufactured and legitimate specimens, and they do so with incredible insight and cunning. This approach to marketing their trilobite inventory has risen to such extremes that it now allegedly includes utilizing state of the art 3D printers to create schizochroal eyes with multilensed detail for inclusion on certain rare, and potentially lucrative, lichid specimens. Such drastic plastic practices are undertaken primarily because Russian dealers know that many supposedly astute collectors in North America, Europe, and Asia use such specific morphological characteristics as a positive sign of a trilobite's quality and authenticity. The only problem is that such lichids never featured this type of compound eye!

"I got into a heated debate with a Russian dealer a few years ago," said Sam Stubbs, a Houston-based lawyer and one of the world's top trilobite collectors. "He was presenting me with an amazing lichid specimen—a type that I had never seen before. I specifically asked him about the eyes, and he handed me a jeweler's loupe so that I could see them more clearly. The rows of lenses on those eyes were clear and beautiful. They convinced me to purchase that trilobite. It was only a few months *after* the show that a scientist told me the species in question did *not* have compound eyes like those."

Considering the escalating interest in their paleontological products, it should be fascinating to see in which direction the Russian trilobite industry travels in the years ahead. Will it continue to flourish, as it has virtually without pause since the crumbling of the Soviet Union? Or will greater academic and collector scrutiny, along with the pressures inherent in competing on the world's complex economic stage, bring it to a grinding halt?

As has seemingly been the case with all aspects of Russia's shift to an increasingly free market economy over the past three-plus decades, St. Petersburg-based trilobite merchants have suffered through the expected degree of growing pains. It is undeniable, however, that despite the occasional controversies generated by their work, their fossiliferous finds continue to both amaze collectors and scientists around the world and focus additional attention on that still often inscrutable land called Russia.

VENTRAL TRILOBITE PRESERVATION

A memorable trilobite hunting expedition took place in upstate New York about 15 years ago. Over the course of a weekend, literally tons of fossil-bearing sedimentary strata were excavated by an eager group of amateur trilobite enthusiasts, each of whom seemed focused on finding the ultimate Paleozoic prize—a perfectly articulated dorsal specimen. Utilizing nothing more than crowbars, sledgehammers, and occasionally their bare hands, these dedicated diggers, ranging in age from 15 to 65, remained intently absorbed in their task—removing sheet after sheet of Ordovician-age rock. The rhythmic sound of metal banging against 450-million-year-old limestone layers could be heard echoing throughout the vicinity as large, briefcase-sized chunks of stratified matrix began flying this way and that until the quarry foreman stopped the proceedings with a rather gruff salutation.

As he raised one freshly extracted rock fragment in the air, he asked "Who tossed this?" in an agitated voice that could be detected in the neighboring county. "I did," came one young digger's rather sheepish reply. "I didn't think we were keeping negatives." Those words served to stoke the quarry boss's festering frustrations to the next level. "Negative?" he shouted. "That's not a negative; that's a ventral trilobite! That's what you're all looking for. All you'd need to do is flip it during preparation and you'd end up with an incredible specimen . . . but *you* don't deserve it!"

Whether or not our unfortunate young digger deserved to keep his foolishly tossed fossil, it is relatively easy to confuse a ventral specimen with a negative example when it comes to trilobites. They may initially look the same, but under closer inspection the differences between the two quickly become apparent. Negative, or counterpart, trilobites are exactly that—the reverse impression in the rock caused by the actual fossilized animal. These examples are often left in the field by overzealous collectors who believe they are of little use, or value, to those looking for "real" trilobite specimens—although, truth be told, bits of shell, delicate genal spines, and even the occasional compound eye can remain stubbornly affixed to a bug's negative side.

In contrast, ventral trilobites are complete examples of the actual fossils that have been preserved and subsequently discovered with the underside of their calcite carapace showing. It has been speculated that certain species of trilobites may have swum upside down during their lives in the primeval seas much like modern horseshoe crabs. Thus, the discovery of various specimens that have become "flipped" within the sedimentary bedding plane isn't particularly surprising. It should also be noted that with minimal effort these inverted examples can be reattached to a proper piece of matrix (that specimen's internal mold counterpart) and dorsally prepared, with the results often producing highly detailed showpieces. If one prefers to prep a ventrally preserved specimen as is, however, these examples often display delicate internal muscle attachment "hooks" and scars, along with occasional appendage remnants positioned along the trilobite's underside. Many ventral specimens also showcase the trilobites' distinctively pronged hypostome, or mouth plate. These calcified, occasionally forked extensions (which possibly allowed the trilobite to attach to a food source, rock surface, or mate) come in a wide variety of shapes and sizes, and when found independently often serve to identify the trilobite species from which they originated.

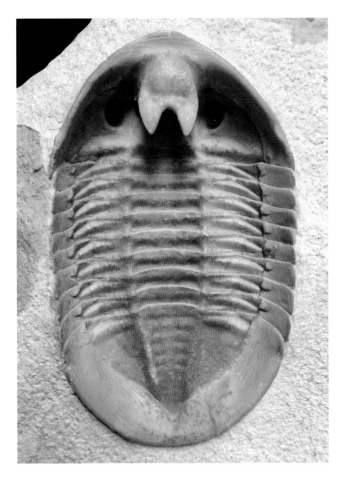

ASAPHUS CORNUTUS (PANDER, 1830)

Middle Ordovician, Llanvirnian; Asery Regional Stage; Duboviki Formation, Vilpovitsky Quarry; St. Petersburg region, Russia; 9.5 cm

The forked hypostome—or mouth plate—is one of the key morphological features exhibited by any properly prepared ventral specimen.

Some locations, such as the famed Burgess Shale in British Columbia, the Hunsruck Slate of Germany, and the Lorraine Shale of upstate New York, have become renowned for producing trilobite ventral examples displaying detailed evidence of soft-tissue preservation, including antennae, claws, gills, and even, on exceptionally rare occasions, eggs. These specimens are not only prized by collectors around the globe but are also the subject of intense study by scientists trying to learn more about the often secretive lifestyle of everyone's favorite ancient arthropod.

WALCOTT/RUST QUARRY, NEW YORK: *WHERE LEGENDS ARE BORN*

In 1870 20-year-old Charles Doolittle Walcott stumbled upon a promising outcrop of Ordovician rock located near the town of Trenton Falls, New York. (Four decades later, he would cement his reputation as one of paleontology's monumental figures with his work at British Columbia's 508-million-year-old Burgess Shale.) In all honesty, it wasn't just by chance that the budding adventurer found himself exploring this exposed streambed on land owned by 44-year-old local farmer William Rust. Indeed, the industrious Walcott had recently purchased a home in the vicinity expressly to facilitate his investigation of the area's fossil-rich sedimentary layers.

Walcott had already spent much of his life searching through the Paleozoic strata of his native New York State, assembling a sizable collection of both minerals and fossils before he even reached his teens. The ambitious young scientist was captivated by the mysteries revealed within these ancient rocks, and each new locale promised additional knowledge, along with the heady mix of potential fame and fortune. By the time Walcott started his own exploration of the Trenton Falls locale, he was well aware that Rust—who had begun excavating his property's rock strata in 1860 primarily to procure building stones—had uncovered fossiliferous layers brimming with trilobites, those ancient arthropods that once dominated the world's oceans. What Walcott perhaps didn't know was that Rust had been gathering the fossils he found scattered about on the family property since childhood, and his renown as an avid collector of these curious remnants of the distant past had made him something of a local celebrity.

With so much in common, it didn't take long for a professional and personal bond to be formed

(strengthened by Walcott soon marrying Rust's younger sister, Lura Ann), and within months of their introduction, the pair began digging together in the quarry's 455-million-year-old limestone layers. During that initial excavation, more than 300 square meters of the hard, fossil-bearing rock was removed by hand and carefully broken down and examined for the telltale signs of partially buried fossils. It was laborious work, enhanced further by the delicate and time-consuming final preparation touch required to fully expose any promising specimens from their surrounding matrix.

The combined efforts of Walcott and Rust revealed a veritable treasure trove of beautifully preserved trilobites. Hundreds of complete examples were unearthed, including such now recognized species as *Ceraurus pleurexanthemus, Isotelus gigas*, and *Bumastoides holei*. In fact, within their first two digging seasons, 18 distinct trilobite species were recovered, along with an equally impressive array of cystoids, crinoids, and brachiopods. Together these intricately preserved Paleozoic relics marked the quarry as one of the most prolific fossil biozones ever found in North America. As the quantity and quality of their discoveries grew, Walcott also began to realize that there was a promising commercial market for these fossils, especially from museums across the Northeast, a few of which eagerly sought to purchase his latest finds. It didn't take long for him to sense that an incredible opportunity—with both scientific and financial ramifications—had fallen his way.

THALEOPS LAURENTIANA AMATI AND WESTROP, 2004

Middle Ordovician; Rust Formation, Trenton Group; Walcott/ Rust Quarry; Trenton Falls, New York, United States; 2.6 cm

One of the rarer faunal representatives of the W/R Quarry, these three-dimensionally preserved specimens present an amazingly lifelike appearance.

"The stories are that Walcott was never shy about promoting himself or his discoveries," said Martin Shugar, a field associate in paleontology with the American Museum of Natural History. "He was reputed to be a very determined, clever, and intelligent man, and he was also among the few at that time to discern that the value of fossils stretched beyond their purely scientific worth."

James Hall, then director of the New York State Museum in Albany, was the first to be approached by Walcott about acquiring a comprehensive collection of material drawn from the quarry. Much to Hall's regret, however, he had to turn down the opportunity, ostensibly due to his institution's limited available funds (although rumors persist that some within the museum's hierarchy found young Walcott's aggressive approach somewhat off-putting). Word of the discoveries next reached the ears of Louis Agassiz, director of Harvard's prestigious Museum of Comparative Zoology (MCZ), and shortly before his death in 1873 Agassiz stepped forward to purchase the best of the Walcott/Rust bonanza for the then exorbitant sum of $3,500 (approximately $70,000 in today's dollars). That imposing fossil assemblage highlighted 325 trilobites, 190 crinoids, and 6 starfish, a good number of which represented both new genera and new species.

Agassiz was so enamored with both the collection and indications that some of the partial trilobite specimens featured soft-tissue preservation (including antennae and limbs), that he encouraged Walcott—both emotionally and financially—to continue his work within the quarry's limestone walls. Due primarily to the backing of Louis Agassiz's son, Alexander—who took over as director of the MCZ after his father's passing—those efforts went on virtually unabated until 1876. During that time, both Walcott and Rust ventured deeper and deeper into the site's fossil-rich layers. While doing so, they managed to uncover and expose the first

HYPODICRANOTUS STRIATULUS (WALCOTT, 1875)

Middle Ordovician; Rust Formation, Trenton Group; Walcott/Rust Quarry; Trenton Falls, New York, United States; 3.5 cm

Found in 2015, this rare specimen is distinguished by huge, crescent-shaped eyes. The species also features a pronounced hypostome (not visible here) that covers more than half the length of its body.

complete trilobite specimens definitively exhibiting appendages, news of which rocked the still nascent paleontological world.

Spurred on by his findings, as well as by growing support from the academic community, Walcott came to realize that he could make a full-time career out of his fossil-centric efforts. As the youthful entrepreneur kept digging within the quarry, he expanded his scientific base by writing several well-received papers focused on trilobite appendages, the first such works of their kind. Over the

next half-decade, Walcott's varied paleontological interests carried him farther and farther afield, especially after the unexpected death of his wife in 1876. Yet no matter where his fossil-inspired ventures led him, Walcott sporadically returned to the quarry, gathering additional material for study and sale, much of which he then promptly moved to either the MCZ at Harvard or to the Peabody Museum at Yale.

Despite his brother-in-law's growing wanderlust, William Rust continued digging and marketing fossils found in the quarry's plentiful layers for another two decades, only stopping shortly before his death in 1897. Through these efforts, he managed to unearth a comprehensive collection of specimens from what eventually became known as the Trenton Group, an imposing geologic unit that appears in Ordovician-age outcrops running between New York and West Virginia. Rust's discoveries represented a Paleozoic assemblage that at that time was perhaps rivaled only by the material amassed by the famed French naturalist Joachim Barrande in Bohemia (now the Czech Republic) a few decades earlier. The New York stratum from which these historic fossils were drawn was named in his honor and is now recognized as the Rust Formation.

"Most people outside of the scientific world tend to overlook the importance of William Rust," said Sam Stubbs, a renowned trilobite collector associated with the Houston Museum of Natural Science. "He kept that quarry going when Walcott began to drift away and pursue other interests."

With the commercial success and scientific acclaim that their efforts had generated, neither Walcott nor Rust could have imagined that nearly a century would pass before further investigative work would take place within the quarry. After so many decades of disregard and decay, the mere existence of their once legendary excavation site had become little more than a rumor to many late-twentieth-century academics. Within the scientific community, most showed barely a passing interest in this long-lost spot in central New York State, and those who chose to focus any of their attention on the region's magnificently preserved Ordovician trilobites contented themselves with studying the impressive array of material already housed within Harvard's hallowed halls.

In 1990, things began to change in and around the Walcott/Rust quarry. After extensive research at the MCZ, and a few get acquainted visits with the current landowner (who had no lingering relationship with the long-departed Rust clan), amateur fossil enthusiast Thomas Whiteley began actively taking steps toward rediscovering and reopening the famed fossil locale. Whiteley believed there was still a great deal to learn from the area's fossil-bearing layers and that modern excavation and preparation techniques could reveal an exciting new story for the famed quarry's trilobites. But first he needed to find it.

Whiteley was not deterred when he learned that sections of the former Rust farm had been converted into a golf course and that a hydroelectric power plant now filled an adjacent gorge he once considered a potential quarry location. But after more than two months of fruitless exploration through the area's thickly forested hills and valleys, his frustrations had started to grow. Finally, just when he was beginning to believe that he might never find the site, some town residents directed him to a promising but now heavily wooded spot adjacent to an algae-covered waterway known as Gray's Brook. There he immediately noted piles of roughly stacked rocks that he believed might be remnants left over from the original nineteenth-century dig. He also noticed a series of interesting sedimentary outcrops that were emerging from an embankment that had since become overgrown with weeds. Once he removed the undergrowth and saw the clear-cut marks of quarrying, he knew what he had been hunting for had been found.

"Finding and then working the quarry was a true passion for me," said Whiteley, who coauthored *Trilobites of New York* in 2002, which featured his photographs of many prime specimens drawn from the Walcott/Rust locale. "I had recently retired from my job at Eastman-Kodak and had come across the original Walcott/Rust collection while taking some trilobite photos at Harvard's museum. The preservation of the trilobites amazed me, and knowing that virtually all of the available material had been collected and prepared over 100 years ago, without the benefit of any modern machinery, convinced me to try and undertake a new dig at the quarry site."

After leasing the land and bringing in an assortment of bulldozers, backhoes, and other excavation equipment the likes of which Walcott and Rust could only have dreamt about, Whiteley set about reopening the quarry within months of his initial "discovery." Before doing so, however, he had to remove several trees and the lush foliage that had overgrown the property during the preceding century. When that task was completed, he began a careful examination of the exposed rock layers in an attempt to determine which were the best candidates to bear fossiliferous fruit. His immediate goal was to find what had become known to Walcott and Rust as the "*Ceraurus* layer"—a thin limestone band that had perfectly fossilized a mass assemblage of those distinctive trilobites, often ventrally, and occasionally with their lightly calcified antennae preserved. The results soon began to exceed even Whiteley's wildest expectations. Layer after carefully excavated layer provided a rich harvest of material, with the trilobites proving to be both abundant and diverse.

Whiteley's work would carry on for the next 15 years. Each season, from late spring until early autumn, he would find time to venture back to the

quarry site to resume his digging activities. Whiteley estimates that he and his small crew moved in excess of 100 tons of rock overburden during that time merely to reach the fossil-bearing layers, which he then explored with patience and fastidious care. Unquestionably, the results he attained were well worth such a backbreaking, time-consuming endeavor. Whiteley found hundreds of complete trilobites during his dig, including rare lichids and illaenids, along with an accompanying array of crinoids, bryozoans, carpoids, and other assorted faunal components.

However, uncovering this nearly half-billion-year-old material was only the first step in his quest; Whiteley also wanted to study these fossils to the best of his somewhat finite abilities. Without direct access to a museum laboratory or university workspace, Whiteley spent much of his nonquarry time at home, polishing trilobite-filled rock cross-sections and acid etching limestone surfaces. In doing this, he exposed never before seen details of trilobite morphology, as well as revealing previously unknown evidence of species density and distribution. He even went back to Harvard to cross-reference his recent finds with those of his illustrious predecessors. Fittingly, when he completed his work at the quarry site in 2005, Whiteley donated virtually all of his discoveries to the MCZ, greatly expanding the museum's original collection obtained from Walcott and Rust over a century earlier.

"The quarry presented such a unique opportunity for study," Whiteley said. "Unlike those who came before and after me, I didn't really have much interest in the commercialization of these finds. Nor was I a professional paleontologist who could effectively interpret and disseminate all the material that was made available by our dig. But I had incredible assistance from scientists like Carlton Brett of the University of Cincinnati, who helped immeasurably in turning this from a mere 'dig' into a truly meaningful scientific mission."

The completion of Whiteley's quarry operations did not signal another prolonged rest for the limestone layers first explored by Charles Walcott and William Rust. In fact, less than two years later an exciting new story began to unfold at the venerable site. Somewhat ironically, during a few of Whiteley's first ventures to the Trenton Falls location in the early 1990s, he had been accompanied by Dan and Jason Cooper, a father-son team of fossil enthusiasts based in Ohio. Both Coopers had gained a great appreciation for New York's sedimentary strata while searching for—and eventually helping to rediscover—the state's equally famed Beecher's Bed of Oneida County, where the first Ordovician-age *Triarthrus eatoni* trilobites with soft-tissue preservation had been found in 1893. The Coopers also realized the immense potential still housed within the rocks of the Walcott/Rust quarry, and upon the completion of Whiteley's work, they quickly stepped in with the intent of leasing the land and undertaking a massive commercial trilobite dig.

Eschewing some of the scientifically inclined subtlety Whiteley had employed during his efforts, in 2008 the Coopers began the most extensive, expansive, and expensive excavation the quarry had yet experienced. Removing layer after layer of sedimentary strata almost exclusively by hand and digging in many of the spots previously cleared by the Whiteley crew, in a single season the Coopers extracted almost as much fossil-bearing rock as their predecessor had managed during the final

ISOTELUS GIGAS (DEKAY, 1824); *CERAURUS PLEUREXANTHEMUS* (GREEN, 1832)

Middle Ordovician; Rust Formation, Trenton Group; Walcott/Rust Quarry; Trenton Falls, New York, United States; 12.5 cm

These large, football-shaped trilobites—occasionally more than 18 centimeters in length—have been discovered in significant numbers at this site. Note that a 3-centimeter *Ceraurus* has hitched a ride atop this impressive specimen.

BUMASTOIDES HOLEI (FOERSTE, 1920)
Middle Ordovician; Rust Formation, Trenton Group; Walcott/Rust Quarry; Trenton Falls, New York, United States; 5 cm

Almost identical examples of this trilobite have been discovered in similarly aged strata in Ontario, Wisconsin, and Minnesota. It is among the more prolific species found in this bucolic New York locale.

most aesthetically appealing trilobites ever found, anywhere. With their thick black shells, intricate surface ornamentation, and incredible, three-dimensional preservation, the trilobites pulled from these rich Ordovician layers were nothing short of spectacular. Some completed specimens were so lifelike in their fossilized appearance that it seemed all one needed to do was drop them in water and watch them crawl away.

The Coopers admit that they weren't particularly surprised when their discoveries made an immediate and dramatic impact on the world's trilobite market. At leading fossil and mineral shows in Tokyo and Tucson, museum officials and private collectors openly sparred with one another in their attempts to acquire the biggest, rarest, and most perfectly preserved trilobite specimens emerging from the famed Walcott/Rust quarry. Huge football shaped examples of *Isotelus gigas* battled for attention while perched next to pristine displays of *Flexicalymene senaria* and *Meadowtownella trentonensis*. Especially rare species, such as the delicately spined, 8-centimeter-long *Apianurus sp.*, went directly into exhibits at the Smithsonian, and other unusual trilo-types, such as the diminutive bubble-nosed *Sphaerocoryphe robusta* and the elegant *Amphilichas cornutus*, ended up in private collections everywhere from Barcelona to New York City to Cancun.

As is so often the case within the highly politicized paleontological world, not everyone was especially thrilled that the revered rocks of the

half-decade of his stay. Their goals were clear: find, prepare, and sell the best trilobites ever uncovered at the famed site, which they firmly believed would rank in quality with *any* trilobites found in *any* other fossil-bearing outcrop on Earth.

They knew their task wouldn't be an easy one. Not only was the hard, freshly excavated limestone rock difficult to dig, it also proved problematic to remove from the trilobites during preparation. But once they mastered the "trick" of effective prep (which in finished quality far exceeded any results achieved by either Walcott or Whiteley), they realized that these indeed were among the

AMPHILICHAS CORNUTUS (CLARKE, 1894)
Middle Ordovician; Rust Formation, Trenton Group; Walcott/Rust Quarry; Trenton Falls, New York, United States; 6.2 cm

Found in 2014, this is only the fourth known complete example of this species—and perhaps the most detailed. Although in many locations lichid trilobites have become flattened during fossilization, those in the W/R Quarry retain much of their lifelike appearance.

Walcott/Rust quarry were now being "desecrated" by the endeavors of commercial diggers. Although some within the fossil community may have looked somewhat askance at the nature of the new dig, Jason Cooper firmly believes that Charles Walcott himself would have wholeheartedly understood and supported their current efforts. After all, Cooper reasons, it wasn't like the legendary paleontologist was against seizing an opportunity when it knocked on his door, especially when it came to adding another exciting and profitable chapter to his beloved quarry's storied history.

"These trilobites deserve to be seen and appreciated," Jason said. "We're finding many of them in sufficient numbers so that both institutions and collectors can obtain perfect specimens. I think we're approaching this dig very much in the manner that Walcott intended—to make sure that key pieces end up in museums, but that everyone is rewarded properly for their work."

TRILOBITE SOFT-TISSUE PRESERVATION

There is a temptation among those who collect and study trilobites to occasionally treat these Paleozoic relics as little more than their fossilized exoskeletons. After all, the pervasive remains of those hard, calcite carapaces represent the most tangible evidence we have confirming the existence of these long-gone marine inhabitants. But there is clearly more to trilobites than just their outer shells. Judging by the results provided in a series of recent scientific investigations, there is *much* more. From their earliest stages in the Lower Cambrian, we know that trilobites possessed calcium-coated eyes—quite spectacular ones at that. And ever since the discoveries made in the famed Burgess Shale over a century ago, we've been aware that trilobite anatomy also featured an impressive array of nonbiomineralized body parts—including

TRIARTHRUS EATONI (HALL, 1838)

Upper Ordovician; Lorraine Shale; Martin Quarry, Beecher's Trilobite Bed; Oneida County, New York, United States; 3 cm

This magnificent pyritized specimen displays antennae, gills, and walking legs among other anatomical features. Hundreds of twenty-first-century finds at this site have been revelatory regarding trilobite morphology.

Photo courtesy of Markus Martin

antennae, gills, multiple walking legs, and even basic respiratory tracts.

Recent finds in the Ordovician strata of upstate New York have provided even more dramatic insight into trilobite morphology, showing clear signs of both eggs and primitive yet apparently highly efficient reproductive systems. And in 2016, fossils unearthed within the 478-million-year-old Fezouata Formation of Morocco supplied even more information about what trilobite internal body design—and subsequent behavior—may have been like in those far distant seas. Working with three exceptionally well-preserved examples of *Megistaspis hammondi*, a rare asaphid species found in Ordovician outcrops located in the southeastern corner of that fossil-rich North African nation, a pair of Spanish scientists—Diego

**MEGISTASPIS (EKERASPIS) HAMMONDI
CORBACHO AND VELA, 2010**

Lower Ordovician, Arenig Series; Upper Fezouata Formation;
Drâa Valley, Zagora area, Morocco; 28.2 cm

It was on ventrally preserved examples of this elegant species
that Spanish scientists discovered previously unknown appendages and internal organs. Unfortunately, that morphology is
not apparent on this dorsal example, although it does display
pronounced antennae.

Garcia-Bellido and Juan Carlos Gutierrez-Marco—
uncovered evidence of a previously unseen mid-
gut gland, which in many modern arthropods is
known to secrete enzymes used for the digestion
of food. Their studies also exposed signs of a crop,
an internal organ commonly found in contemporary sediment feeders, especially those that live
along the seafloor.

These discoveries presented seemingly irrefutable proof supporting the long-held belief that at
least some species of trilobites spent their lives foraging through the ocean bottom's organism-rich
mud. And judging by the size of these Moroccan
Megistaspis trilobites—occasionally up to 35 centimeters in length, including a long, graceful tail
spine—it appears that a bottom-dwelling lifestyle
suited both their needs and nature remarkably
well. These recent findings also noted three pairs
of short, powerful appendages located on the
underside of each specimen's head—the first time
such an anatomical feature has been detected in a
trilobite fossil. These cephalic limbs were particularly eye-opening to academics because they differ
radically in both form and apparent function from
the longer, smoother walking legs that adorn the
underside of most trilobite thoraces and pygidia.

The Spanish scientists speculated that these
newly noted appendages may also have been
responsible for creating the trace fossils known as
Cruziana rugosa, or trilobite trackways. It had long
been surmised that the back-and-forth movement
of certain trilobite walking legs were the cause of
these mysterious trace fossils, which for centuries
have been found in conjunction with trilobites in
sedimentary deposits around the globe. These ventral *Megistaspis hammondi* discoveries supplied
the corroborative evidence that paleontologists
had long sought to confirm their theories. Despite
this compelling new information, it is generally
assumed that not all trilobites were sediment-feed-
ing bottom dwellers like *Megistaspis.*

It has been theorized that during their 270-mil-
lion-year existence trilobites successfully inhabited virtually every available environmental niche
in their ever-changing water world. Some were
"floaters," able to effectively ride upon sun-driven
ocean thermals; and others were active swimmers,
darting in and out of primal crinoid beds and coral
reefs. Still others may well have been the scourges

of the early seas, coasting along on strong ocean currents and preying on soft-bodied arthropods and on smaller trilobites. However, despite the exciting evidence supplied by the recent Moroccan finds, definitive proof for such trilobite lifestyle diversity has yet to be uncovered within the fossil record. Perhaps one day soon, in some distant corner of the globe, we will unearth specimens that provide additional insight into trilobite ventral, soft-tissue anatomy, discoveries that will further demonstrate the unique adaptive abilities of these amazing creatures that inhabited the ancient seas half-a-billion years ago.

FILLMORE FORMATION, UTAH:
AMERICA'S TRILOBITE PARADISE

Trilobites can be found just about anywhere. In the United States alone, fossil-filled deposits run the Paleozoic gamut from Lower Cambrian to Upper Permian, with many of these sedimentary sites containing examples of widely varying trilobite genera and species. These amazingly adaptable arthropods have been uncovered from New York to California, and from Minnesota to Texas, with a preponderance of the states in between reporting either an abundance of faunal material or perhaps just an occasional find or two. Although Pennsylvania, Ohio, Wisconsin, Kansas, Georgia, and Nevada may lay claim to being the most bountiful trilobite repositories in the country, the truth is that nowhere across the expanse of the entire North American continent are trilobites more prevalent than in the majestic state of Utah.

This mountainous western outpost brims with some of the most renowned and studied Paleozoic horizons in the world. The state's geologic exposures range from the Middle Cambrian Wheeler, Marjum, and Weeks formations, through the Lower Ordovician Fillmore and Wahwah outcroppings, with a few rarely explored and poorly

preserved Lower Devonian deposits thrown in for good measure. Incredibly, more than 500 scientifically recognized trilobite species have been described from Utah's fossil-rich sedimentary strata, with many being indigenous only to this region's densely packed limestone layers. After more than a century and a half of investigation and analysis, new species are still being unearthed on a regular basis, most by adventurers who have been literally and figuratively willing to travel that extra step for the opportunity to dig deep into some of the state's previously untapped fossiliferous reserves.

Trilobite collecting in Utah enjoys a long and noble history. Going back at least to the eighteenth century—and most likely much earlier—members of the native Ute tribe found strange "animals of stone" lying on the ground (the calcified shells of 2- to 5-centimeter-long *Elrathia kingii* trilobites), drilled holes into their pygidia, and wore these half-billion-year-old fossils around their necks as good luck amulets. To further magnify the perceived protective powers of these impressive invertebrates, Native American petroglyphs depicting what clearly appear to be trilobites have been found adorning ancient cliff faces throughout the region. The Ute may have been the first to recognize the state's trilobite fauna as something extraordinary, but they certainly weren't the last.

Among the prominent names that subsequently searched for these primordial relics throughout Utah's rugged landscape was the seemingly omnipresent Charles Walcott. During his 50-year

PRESBYNILEUS IBEXENSIS (HINTZE, 1952)
Lower Ordovician; Fillmore Formation, Pogonip Group; Millard County, Utah, United States; 8.1 cm

This is one of the most prevalent trilobites found within the Fillmore Formation. Once scarce, excavations over the last two decades have made this species an expected component of most private collections.

TRIGONOCERCA PIOCHENSIS (HINTZE, 1953)

Lower Ordovician, Blackhillsian Stage; Fillmore Formation, Carrolinites nevadensis Zone; Southern House Range, Millard County, Utah, United States; 10 cm

---

This may represent the largest known articulated specimen of this hard to find species.

career, he played an essential role in the discovery and exploration of both British Columbia's famed Middle Cambrian Burgess Shale and New York State's renowned Ordovician-age Walcott/Rust quarry. Prior to taking the prestigious "desk job" as secretary of the Smithsonian Institution in 1907,

ISOTELOIDES FLEXUS (HINTZE, 1952)

Lower Ordovician; Fillmore Formation, Pogonip Group; Ibex Springs, Utah, United States; 4.2 cm

---

Disarticulated fragments of this elegant trilobite indicate that some examples grew to prodigious dimensions, perhaps up to 40 centimeters in length.

Walcott frequently visited Utah in search of new trilobite-bearing exposures. He was first to name both the Wheeler and Marjum formations, noting the "alternating bands of thin-bedded limestone and calcareous shale" that characterized these fossil-filled outcrops. But it wasn't until the pioneering work of local resident Lloyd Gunther and his family in the mid-1960s that the state's Middle Cambrian trilobite bonanza began to be fully explored and recognized. The Gunthers' efforts helped discover and identify complete examples of such species as *Hemirhodon amplipyge*, which could reach 15 centimeters in length, and *Bolaspidella housensis*, a diminutive trilo-type that rarely exceeded a centimeter in size.

The work of the Gunther clan has now continued unabated for more than half a century, with new generations of the family still working these central Utah layers (and others throughout the state) on a regular basis. Lloyd Gunther's son, Val, served as coauthor of the well-received *Exceptional Cambrian Fossils from Utah* in 2015, which for the first time identified many of the state's rarer trilobite species. And grandson Glade has become an outspoken opponent of government-imposed restrictions on fossil collecting within many of the state's key Paleozoic sites—a troubling trend that has continued to escalate throughout the twenty-first century.

"My entire family has been involved with finding trilobites in Utah for as long as I can remember," Val Gunther stated. "During his lifetime, my dad found and donated hundreds of specimens to universities and museums, and it was my mom who found the first complete *Hemirhodon* back in 1967."

The Gunthers are far from alone in their ongoing appreciation and pursuit of Utah's amazing array of trilobites. In recent years, fresh waves of explorers, academics, and collectors—all focused on procuring previously unseen species from previously

LEMUREOPS LEMUREI (HINTZE, 1953)

Lower Ordovician; Fillmore Formation, Pogonip Group; Millard County, Utah, United States; 2.4 cm

Recently reclassified from *Pseudocybele*, this is one of the hallmark species of the Fillmore Formation.

Photo courtesy of the M. Shugar Collection

unknown horizons—continue to be drawn to the state's fossil riches on an annual basis. Among the more notable of this new breed of enthusiasts is Terry Abbott, a full-time commercial fossil collector based in the small Utah town of Delta, who realizes how fortunate he is to be living amid some of the most famous and abundant trilobite outcrops on the planet.

"What better place could there be than Utah for someone interested in trilobites?" Abbott asked with a smile. "It's almost impossible to travel more than 30 minutes from my home in any direction and not find a great digging spot. Of course, it does help to know where to look."

Not only does Abbott revel in both the diversity and availability of Utah's trilobites but also in the geologic history of the state itself. He is well aware that in the Middle Cambrian, some 510-million-years-ago, this now often arid inland region was part of a great marine shelf adjacent to the super-continent of Laurentia, which then lay near Earth's equator. In those tropical seas, amid myriad coral-generated reefs and swaying fields of ancestral seaweed, early life—including an ever-expanding variety of trilobite species—proliferated at a rate unprecedented in the planet's four-billion-year history.

At that long distant time in Earth's past, Utah's now mountainous terrain served as the floor of a shallow sea, and early denizens of the deep existed in what was essentially an offshore sanctuary. Tri-lobites of all imaginable (and unimaginable) types swam through these waters, along with an abun-dance of soft-bodied arthropods that in their mul-tiplicity rivaled those found in any other Cambrian location on the planet. Three orders of trilobites, featuring scores of genera and hundreds of dis-tinct species (such as *Modocia typicalis*, *Olenoides nevadensis*, and *Bathyuriscus fimbriatus*), have been identified from the state's Cambrian layers alone, along with a variety of worms, lobopods, and giant arthropods. The latter group included "sea monsters" like *Anomalocaris canadensis* and *Beckwithia typa* that reached 25 centimeters in length and most likely feasted on the diverse fauna (including trilobites) that shared their marine world.

Upon their demise, a significant number of tri-lobites ended up as part of an ecosystem lurking at the bottom of this fertile ocean environment. Their calcium-coated remains—including an aggrega-tion of their frequently molted and often disartic-ulated exoskeletons—were slowly covered by layer upon layer of mineral laden sea sediment, which over the ensuing eons formed hard limestone and sandstone bands that half-a-billion years later would produce some of the most prolific Paleozoic fossil repositories ever discovered.

"You can go into places like the *Elrathia* beds and find hundreds of trilobites in a day," Abbott said. "A lot of them are molts, but that's to be expected. But those kinds of deposits are unusual; most of the time it takes a lot of effort to find just a single specimen."

Over the past three decades, Abbott has explored virtually all of Utah's renowned Middle Cambrian trilobite localities. Many of his adven-tures have carried him to sites within the House Range, the imposing mountain chain that includes both the Marjum Formation (home to such rare collector's favorites as *Olenoides pugio* and *Glosso-pleura gigantea*) and the slightly younger Wheeler Shale, where pristine examples of the abundant species *Elrathia kingii* and *Asaphiscus wheeleri* have become staples in every beginner's fossil collection.

More recently, Abbott's trilobite-related travels have brought him in direct contact with a vari-ety of exciting, formerly uncharted fossiliferous formations, most of which are situated far off the beaten track. Amid these remote exposures, he has begun uncovering a spectacular array of complete Cambrian and Ordovician trilobites—some previ-ously unknown to science. Other species emerg-ing from these excavations had been recognized earlier only through disarticulated bits and pieces collected over the years by university academics who found them during brief and often superficial surface explorations.

Abbott's groundbreaking work, traditionally undertaken by hand without the benefit of motor-ized rock moving machinery, has provided both form and substance to many of these hitherto frag-mentarily known trilo-types. Of particular note is

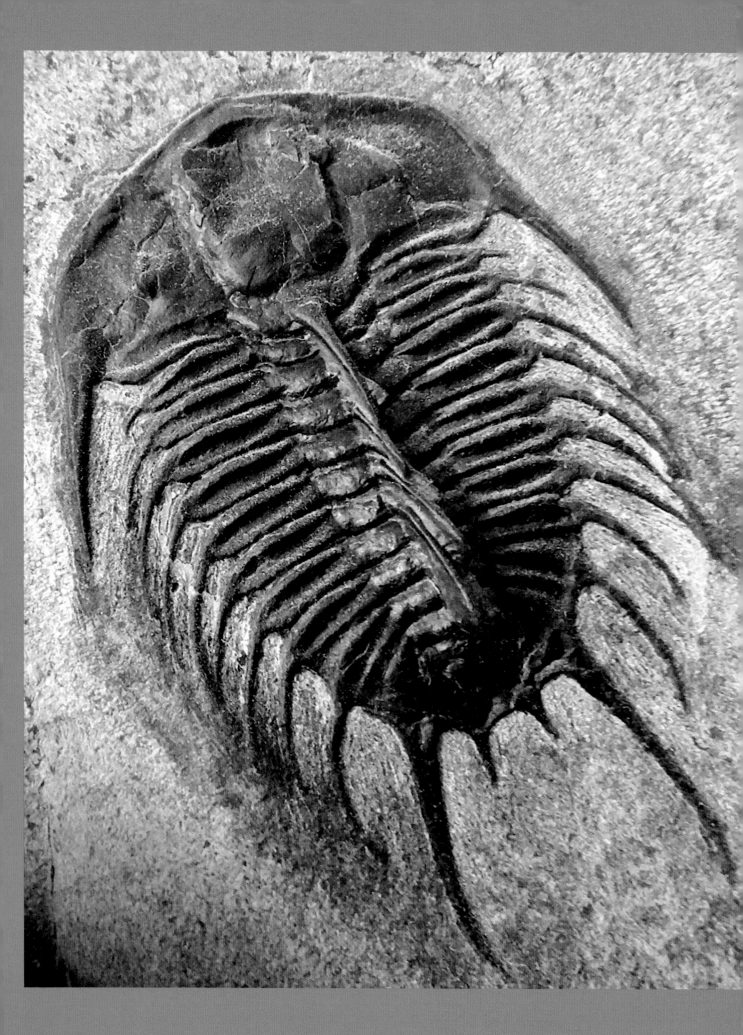

a diverse assortment of trilobites unearthed from the 460-million-year-old Lower Ordovician Fillmore Formation. Within this fossil-rich limestone layer, the molted shells of larger specimens (that judging by the frequently shattered and scattered evidence could have attained sizes of 25 centimeters or more) were apparently torn asunder by near constant subsurface storms, leaving only smaller complete examples (usually 5 centimeters or less) to be revealed.

Abbott is quick to downplay both the scientific significance and the potential commercial windfall generated by his work, preferring to focus more on the pure excitement of discovery. But even after spending most of his adult life in pursuit of these often elusive fossils—with his journeys repeatedly carrying him away from his family and toward a dot on the map destination located deep in the heart of Utah's convoluted maze of canyons and arroyos—his hunger for making new finds remains as strong as ever. Some academics within the trilobite world may look askance at the efforts of a so-called commercial collector, but Abbott has been continually surprised by the positive response his work has received from key members of the scientific community.

"Professor Dick Robison from the University of Kansas has been out digging with me a few times," Abbott said. "He's about the only person, other than my wife, who I've allowed into my latest Ordovician site. I always let him keep the first complete specimens we find, and I hope they prove to be important to his future studies. Over the years, I've come to believe that the diggers, the scientists, and the collectors should try to work together whenever possible."

Abbott's good-natured attitude toward gathering and distributing his fossil material has played a major role in yielding a bumper crop of trilobite "trophies," both for himself and for his scientific comrades. Among his more intriguing finds over the last few years are complete examples of such unusual Ordovician species as *Hintzia aemula, Trigonocerca piochensis, Presbynelius ibexensis,* and *Ptyocephalus yersini*—all previously known to science but rarely, if ever, as articulated specimens. His Cambrian discoveries have included *Utaspis marjumensis, Tricrepicephalus coria,* and *Olenoides superbus* (distinguished by spines running down its axial lobe) as well as a new and so far undescribed species of *Modocia* featuring strange bulbous nodes emanating from each of its thoracic segments.

Of course, not every specimen Abbott uncovers is rare or perfect. He admits to finding examples of incomplete or more common species at approximately a 100:1 ratio to his more exotic discoveries. Apparently, patience and persistence are two virtues a good trilobite collector must have in abundance during field excursions. Abbott reveals that a good day's work under the unforgiving Utah summer sun may yield only one or perhaps two complete specimens, and he can recall too many occasions when he made the three-hour drive home with nary a "keeper" to show for his backbreaking efforts. Somewhat surprisingly, this is one paleontological prospector who seems to derive almost as much pleasure from digging up a relatively well-known species (especially if its fossilized carapace is in good condition) as he does from unearthing a one-of-a-kind trilobite treasure.

"If your only goal is to find unique specimens, or previously unknown species, you're going to be disappointed almost all the time," Abbott explained. "I'm not saying I don't love finding

OLENOIDES NEVADENSIS (MEEK, 1870)

Middle Cambrian; Wheeler Formation; Utah, United States; 12.2 cm

This is perhaps the most coveted Cambrian trilobite drawn from the fossil-rich Utah Lagerstatte. It is only over the last two decades that modern prepping techniques were able to fully display the characteristic profusion of axial spines on this species.

OLENOIDES PUGIO (WALCOTT, 1908)

Middle Cambrian; Marjum Formation; Utah, United States; 9.3 cm

This ranks among the more unusual examples of a highly success-ful genus. Differing *Olenoides* species have been found in west-ern Canada, China, and Siberia, as well as throughout Utah.

special trilobites, but believe it or not, I get almost an equal thrill from each trilobite I find."

Abbott has yet to uncover an adequate number of the state's rarest Lower Ordovician trilobite species to make them commercially available to collectors, although doing so remains one of his primary fossil driven ambitions. Since he started his Ordovician explorations in 1996, he has marketed approximately two dozen complete specimens (both enrolled and outstretched) of the small pliomerid *Lemureops lemurei* to enthusiasts around the globe. And recently he found a pocket of the midsized asaphid *Isoteloides flexus*, examples of which he plans to prepare and sell to his

contacts over the next few years. Abbott acknowledges that he would *much* rather spend time in the field charting new fossil locations and unearthing potentially unknown trilobites than "wasting my energy" marketing his finds, an activity he views as little more than an economic necessity.

In the future, he hopes to discover enough samples of such scarce Ordovician trilobite species as *Bathyurellus teretus* and *Kanoshia kenoshensis* to make them readily available to both his scientific associates and leading members of the collecting community. Before he can do that, however, he's going to have to spend even more time in the field looking for the perfect digging locale on land he then intends to lease from the state of Utah. Finding prime sedimentary outcrops, Abbott confesses, is quite often just as much a matter of luck as skill. But he also readily admits to having a few time-tested tricks still hidden up his well-worn sleeve, each of which has, at one time or another, greatly aided in his ongoing quest to uncover the ideal collecting site.

"I go back and read some of the geology journals that were written many years ago," he said. "In some cases, those give me an idea where I should look for new trilobite locations—or at least some locations that haven't been explored in a long time. It's amazing to me that in many of those places the original visitors only did surface collecting; they never even bothered to dig into the rock layers. To me, doing the backbreaking dirty work is the *only* way of finding good, complete specimens."

In light of recent actions taken by the United States Congress, Abbott is aware of where he can, and where he *cannot*, break rock. The Omnibus Land Management Act of 2009 has further limited his collecting options, making the digging of any fossil on certain designated parcels of federal land a potential offense. Abbott has procured a variety of leases from local authorities over the years, but some of his favorite Utah trilobite locations now fall squarely on federal property, areas strictly controlled by the Bureau of Land Management (BLM). Although admitting his frustration at what he perceives as increased and unnecessary government intervention, Abbott has reacted by redirecting his collecting energies to state-controlled land (where collecting a limited number of invertebrate fossils is still legal) and privately owned quarries.

"I don't let too many things get in my way, including the BLM," he said with a sly grin. "All I want to do is what I love, which is digging for trilobites . . . while obeying the law."

Each morning when the bright Utah sun reaches over the nearby Drum Mountains and begins to shed its light on the surrounding hillsides, Abbott's thoughts turn toward hopping into his four-wheel drive vehicle and getting his hands dirty out in the field. He believes that each day could be *the* day when he finds the trilobite mother lode—the place where a digger's dreams are made. And whether he uncovers a single trilobite that day or 20, he knows that his thirst to find these special remnants of Earth's distant past will never be fully quenched. Collectors like Terry Abbott hold an unyielding conviction that there will *always* be a bigger, better, rarer specimen waiting for them just under that next outcrop of rock.

TRILOBITE EGGS

Hen's teeth. Unicorn droppings. Trilobite eggs. All are apparently the stuff of myth, mirth, and legend. However, before we place each of these esteemed "rare as" punch lines into a single derisive category, let it be stated loudly and proudly that at least one of them—and unquestionably the one that would most intrigue those perusing these pages—is very much the real deal. Make no mistake about it, trilobite eggs did indeed exist! Evidence of trilobite eggs has long been sought but had never been verified in the fossil record . . . at least until very recently.

Some scientists had even referred to any such irrefutable indicator of trilobite reproductive behavior as the Holy Grail of invertebrate paleontology.

The famed French naturalist Joachim Barrande, lauded for his work with Bohemian trilobites, cited the subject of trilobite eggs as far back as 1872. The legendary Charles Walcott, perhaps best known for his exploration of British Columbia's Burgess Shale, spent much of his later career searching for signs of trilobite eggs amid the myriad specimens he unearthed during his fossil-filled excursions across the face of North America. Walcott was confident that fossilized evidence existed of these primal arthropod ova, but despite extensive efforts to uncover such elusive affirmations of trilobite procreative practices, he was never able to find definitive proof of their presence. Now, some 150 years after Barrande and Walcott made their initial forays into trilobite egg research, an academic study (conducted and written by the paleontologist Thomas Hegna and renowned trilobite hunter/researcher/preparator Markus Martin) has finally offered unassailable confirmation of the long-standing beliefs shared by these scientific stalwarts.

As seen in a small number of carefully prepared examples of *Triarthrus eatoni* uncovered in the Ordovician-age Lorraine Shale of Oneida County, New York, it is evident that trilobites did reproduce through the deposition of eggs. These beautifully fossilized 450-million-year-old trilobites, found amid the long-renowned and often studied location known as Beecher's Trilobite Bed (named after famed Yale University paleontologist Charles Emerson Beecher, who first collected at the site late in the nineteenth century), are typically preserved in a vivid golden pyrite that accentuates their appearance against the surrounding charcoal-gray matrix. These eye-catching, 1- to 3-centimeter-long *Triarthrus* trilobites have garnered worldwide appreciation not only for their

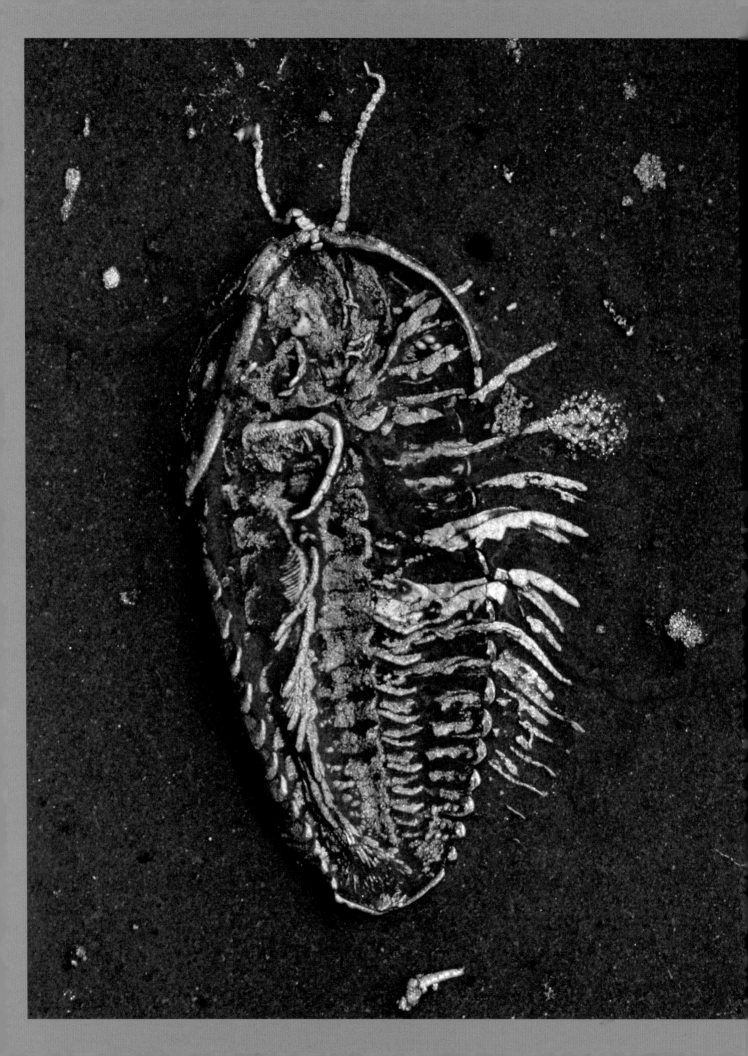

distinctive manner of fossilization but also for their unique soft-tissue preservation, which frequently includes intricately detailed gills, claws, and antennae. Yet despite recovery of thousands of complete specimens from this locale over the last century-plus, only in recent years has the unmistakable evidence of trilobite eggs (some appearing as multicell zygotes) been found and confirmed.

The trilobite eggs discovered within a precious few of these pyritized examples are easy to overlook, appearing as little more than diminutive, circular, golden dots on the dark matrix. Even under high magnification they are difficult to distinguish for all but the most highly trained eye. The 2017 scientific paper that first addressed the subject indicates that these eggs are usually found within the cephalic (head) cavity of ventrally preserved *Triarthrus* specimens and are often grouped in tightly packed clusters, much like the eggs of many living arthropods. These landmark discoveries are the direct by-product of a trilobite bounty unearthed by Martin during a twenty-first-century excavation that reopened the "long lost" site of Beecher's earlier dig. In the process of his explorations, Martin managed to recover hundreds of complete *Triarthrus* specimens as well as a small number of the trinucleid trilobite, *Cryptolithus*, which featured soft-tissue preservation. Thanks to modern preparation techniques, many of these specimens subsequently revealed previously unseen, and unknown, aspects of early arthropod anatomy.

Discovery of Ordovician-age trilobite eggs certainly ranks as an amazing scientific breakthrough, but the finds made at Beecher's Bed don't necessarily represent the oldest eggs ever found in the fossil record. Indeed, some controversial evidence exists for Cambrian trilobite eggs. And in 2004, Chinese scientists working in the Doushantou Formation of South China found clear evidence of 600-million-year-old eggs deposited by what is believed to be a primitive coral-like organism. But most of us would probably agree that tiny, coral-related creatures are a far cry from trilobites. Although the concept that trilobites most likely produced eggs has long been an accepted part of paleontological thought, recent evidence from the rich Ordovician deposits of upstate New York somehow make these ancient arthropods even more appealing and accessible both to those who study and those who appreciate these long-departed remnants of the Paleozoic seas.

VALONGO FORMATION, PORTUGAL: *LAND OF THE GIANTS*

Near the small Portuguese town of Arouca sits one of the most fascinating Ordovician-age trilobite localities in the world, the Louzeiras de Canelas quarry. The invertebrate fossils strikingly showcased within this site's smooth slate blocks are not particularly unusual. In fact, virtually all the species found here since the beginning of the twenty-first century have been previously reported from similarly aged Paleozoic zones throughout France, Spain, and the Czech Republic. Nor is the preservation of the trilobites unearthed within these 450-million-year-old deposits especially remarkable; their flat, often metamorphosed, ghostly white internal molds—some occasionally dusted with golden pyrite crystals—contrast vividly against a charcoal-gray mudstone matrix. But something makes the ancient arthropods drawn from this area's Valongo Formation layers of pronounced scientific significance—the apparent dimensions these trilobites once attained.

TRIARTHRUS EATONI (HALL, 1838)
Upper Ordovician; Lorraine Shale; Martin Quarry, Beecher's Trilobite Bed; Oneida County, New York, United States; 0.9 cm

A twenty-first-century dig at this famed location has yielded dozens of intricately detailed specimens, many with their soft parts (including legs and gills) preserved in pyrite.

Photo courtesy of Markus Martin

ECTILLAENUS GIGANTEUS (BURMEISTER, 1843)
Ordovician; Valongo Formation; Louzerias de Canelas Quarry; Portugal; 12.1 cm

This specimen exhibits the Canelas Quarry's "classic" ghostly white preservation.

Some of the quarry's species, such as *Ogyginus forteyi, Hungioides bohemicus,* and *Uralichas guitierrezi,* reach lengths of 40 centimeters or more, making these primeval treasures among the largest trilobite fossils found anywhere on Earth. Indeed, one partially folded *Ogyginus* specimen has been estimated to have been 72 centimeters long . . . a world record! The trilobites found within these Valongo layers have not only proven to be impressive in size but have also shown themselves to be surprisingly abundant. In the spring of 2009, the revelation that clusters of supersized trilobite fossils had been discovered in these hilly Portuguese outcrops generated a major academic stir. Debates discussing everything from possible mating assemblages to mass mortality scenarios dotted paleontological literature and dominated science-oriented websites. These often pithy conversations embraced such highbrow topics as primordial planetwide climatic shifts, Paleozoic predator-prey interactions, and the role played by fluctuating Ordovician ocean levels in dictating the size variance of the period's marine fauna. It was all enough to make any trilobite enthusiast's head spin!

Mainstream media outlets even appeared to briefly enjoy jumping aboard this trilobite-inspired excursion into the Deep Time past. Within weeks of the discovery's announcement, reporters from local TV stations, globe-spanning podcasts, and international newspaper syndicates all made pilgrimages to the Canelas quarry site in their attempts to better spread this timeless tale to a curious world. This unexpected flood of press attention directed toward the Arouca fossil finds did a great deal to promote the concept that these nearly half-billion-year-old creatures were more than mere mineralized impressions left upon some serendipitous sedimentary surface. Media reports—often accompanied by photos of smiling quarry visitors kneeling next to extracted blocks covered in oversized trilobite remains—seemed to bring these long-gone arthropods back to life, providing previously unknown and unimagined insight into both trilobite morphology and behavior.

NOBILIASAPHUS DELESSEI (DUFET, 1875)
Middle Ordovician, Llanvirn Series; Valongo Formation; Louzerias de Canelas Quarry; Valongo, Portugal; 16.8 cm

This species could reach prodigious dimensions, with some more than 40 centimeters in length. The Valongo Formation has become renowned throughout the world for the unusual size that some of its trilobites attained.

HUNGIOIDES BOHEMICUS AROUQUENSIS (THADEU, 1956)

Lower Ordovician; Valongo Formation; Louzerias de Canelas Quarry; Arouca, Portugal; 41 cm

This is one of the largest and rarest species found amid the oft-oversized Valongo fauna. A similar species has been unearthed in China.

It's easy to understand why these impressive Valongo Formation trilobite assemblages struck a responsive chord in so many of those who saw them. Some stone slabs featured hauntingly preserved images of rare species such as *Placoparia tournemini* and *Eoharpes cristatus*. Others presented nearly a dozen large asaphid trilobites lying virtually one atop the other, with their fossilized carapaces filling the surrounding matrix with unmistakable evidence of their long-ago lives.

In recognition of the myriad trilobite specimens found in the Canelas quarry, the entire region surrounding these shale layers in northern Portugal has recently been designated a UNESCO sponsored Geopark. Since formation of the Geopark concept in 1998, more than 165 sites in 44 nations have been so acknowledged, specifically to assist in preservation of the world's most notable geological resources. This fossil-rich outcrop of the Valongo Formation has now earned global recognition as home not only to some of the largest trilobites in the world but also to some of the most scientifically significant.

"It is without question one of the most interesting paleontological discoveries made on the Iberian Peninsula over the last few decades," said Francisco Alonso Couce, a resident of Madrid who specializes in collecting the giant trilobites of the Valongo Formation. "Finding these huge trilobites was very unexpected . . . especially in the numbers that have been recovered."

Generally, when one thinks about trilobites, images of small, hard-shelled animals scurrying across ancient seafloors instantly spring to mind. And for the most part, these perceptions aren't far removed from the Paleozoic reality. The clear majority of the more than 25,000 scientifically recognized species that existed during the 270-million-year trilobite reign from the dawn of the Cambrian to the end of the Permian were diminutive (7 centimeters or less in length). Indeed, among scientists and collectors, trilobites larger than 10 centimeters are generally considered behemoths. But by no means should such information be interpreted as indicating that trilobites didn't routinely grow larger than that. In fact, the rock-hard evidence provided by a relatively recent deluge of fossiliferous material from around the globe indicates that trilobites actually grew *much* larger! Among the more renowned of these supersized trilobite species have long been the *Isotelus maximus* specimens drawn from Ordovician-age outcrops found throughout the midwestern United States. These imposing organisms often grew up to 35 centimeters in length, and after nearly a century of being collected and admired, they were officially recognized as the State Fossil of Ohio in 1985.

In addition, over the past 30-plus years, a stunning variety of colossal trilobite species have been found in the prolific Ordovician quarries that border St. Petersburg, Russia. On rare occasions, these formidable examples, which include various members of the *Rhinoferus* genus, have exceed 25 centimeters from head to tail. Also, impressively sized Silurian-age trilobite fossils have been uncovered in New York State since the early 1800s, with species such as the unusual homalonotid *Trimerus delphinocephalus* attaining lengths up to 20 centimeters.

During the past hundred years, partial specimens (usually disarticulated pygidia) have provided ample evidence of the great size some trilobites may have attained—such as the legendary spinose Devonian lichid *Terataspis grandis*, which apparently grew up to 50 centimeters in length. Despite such tantalizing clues, complete examples of these extraordinary discoveries have generally been few and far between, at least until the fossil revolution that has occurred worldwide over the past three decades.

Prior to the early 1990s, when the fossil hotbeds of Morocco, China, and Russia initially opened their national doors to the prying eyes of eager scientists, museum curators, and serious collectors, most trilobites had been found in eastern Canada, the western United States, and Central Europe. These generally medium-sized specimens rarely exceeded 8 centimeters in length. Few within the academic and collecting communities had any good reason to suspect that a veritable bonanza of giant-sized Paleozoic relics was out there, just waiting to be unearthed and placed in their fossil-loving hands. But then strange things started to happen.

Late in the twentieth century, the now legendary trilobite beds of northwest Africa started to produce Cambrian-age *Paradoxides* specimens of prodigious size . . . and in prodigious numbers. Dozens upon dozens of complete examples, some reaching 40 centimeters in length, began to emerge on the world stage, highlighting museum displays, flooding fossil trade shows, and even appearing in local rock shops. These massive, 500-million-year-old creatures almost single handedly changed the public perception of trilobites. No longer were these long-gone arthropods viewed as silver dollar–sized fossils that could neatly fit within well-manicured cabinets of curiosities or

on carefully curated museum shelves. These often imposing monstrosities could now be seen from across the room, even if that room happened to be a grand gallery filled with hundreds of invited guests.

It rapidly became clear that many parts of the Moroccan Paleozoic Lagerstatte had served as a final resting place for megafauna. Indeed, that nation soon proved to be crisscrossed by a series of singularly distinctive sedimentary stratum ranging from Lower Cambrian to Upper Devonian and filled with some of the most astounding trilobite specimens ever seen. The appearance of these grandly sized North African arthropods made many within the scientific community begin to reconsider their views concerning trilobite dimensions and lifestyles, but they also made certain scholars ponder why other Paleozoic repositories around the globe weren't routinely producing similarly gigantic fossil specimens.

Some of their questions were soon answered. At roughly the same time that these hulking Moroccan trilobites were hitting the world market, 20-centimeter-long examples of the Ordovician dikelocephalid *Asaphopsoides brevica* were emerging in impressive numbers from China's Hunan Province. In 1998, along the shores of Hudson Bay in Manitoba, Canada, a 71-centimeter-long specimen of the Ordovician asaphid *Isotelus rex* (roughly King of Trilobites) was being collected by a field party co-led by the Manitoba Museum and the University of Manitoba, making it the largest complete, outstretched trilobite yet found.

URALICHAS AFF. GUITIERREZI (RABANO, 1989)
Middle Ordovician; Valongo Formation; Louzerias de Canelas Quarry; Arouca, Portugal; 16.5 cm

This trilobite illustrates the negative side of one of the most renowned discoveries made at the quarry. The dorsal specimen now resides in a prominent place in the town of Arouca's on-site museum.

A decade later, teams of commercial diggers working in Arouca's Canelas quarry—who had been supplying town residents with flat plates of black Valongo Formation slate for use as roofing and paving tiles for more than a century—started to notice that they had hit upon an interesting and somewhat fortuitous layer. Indeed, that 450-million-year-old fossiliferous wellspring had begun yielding an incredible variety of large trilobites, the likes of which had never previously been seen in the region. At first, Portuguese quarry masters looked upon these "intruders" appearing upon their mudstone blocks as little more than an unexpected nuisance. But as soon as both local and international scientists became aware of these amazing arthropods, they flocked to the site by the carload. Their objective was simple—make sure that these extra-large Ordovician trilobites were not only preserved for posterity but also subjected to immediate and intense study. When this academic contingent started to compare these unanticipated Valongo discoveries to the oversized fossil finds made around the globe during the previous few decades, an obvious question began to emerge: Why did trilobites in certain locations grow to astounding dimensions, whereas similar species in bordering outcrops remained within more conventional size boundaries? It was a query that quickly got paleontological tongues wagging from Lisbon to Los Angeles.

One hypothesis speculated that arthropod gigantism could be triggered as an adaption to colder marine climates. In the mid-Ordovician, the undersea location of present-day Portugal was located significantly south of the equator, so it was possible that such gargantuan Arouca trilobites as *Ectillaenus giganteus* and *Asaphellus toledanus* grew larger to meet new environmental pressures placed on them by the changing ecosystem. Other scientists were quick to point out that during that same Paleozoic time frame present-day Hudson

**OGYGINUS FORTEYI
(RABANO, 1989)**

**Lower Ordovician; Valongo Formation;
Louzerias de Canelas Quarry; Arouca,
Portugal; 37.4 cm**

One of the Canelas Quarry's most prolific finds, this species has been named
in honor of the famed British paleontologist, Richard Fortey.

Bay, where *Isotelus rex* was found, was then believed to be part of a warm, subtropical continental shelf. Thus, the debate rages on with vociferous—and confident—supporters on both sides of the discussion.

"It is possible that factors such as water temperature and depth may have played a role in determining the size variance of trilobites," said Sam Stubbs of the Houston Museum of Natural Science. "But so could any number of other circumstances, including the number of predators in a certain habitat."

Prior to the discoveries at the Arouca site, parties specifically search-ing for Paleozoic relics on the western Iberian Peninsula traditionally found these trilobites in pockets featuring a rough, light-gray limestone matrix. Common Valongo Formation species such as *Nesuretus tristani*, *Zeliszkella toledana*, and *Bathycheilus castiliani* have become familiar to both academics and trilobite hobbyists around the globe. But these examples were all relatively small, usually not exceeding 8 centimeters in length, and were often preserved three-dimensionally, with a yellow mineralized coating. Thus, it wasn't until the commercial tile quarry in Arouca started producing its amazing fossil discoveries early in the twenty-first century that a greater appreciation of the possible dimensions that these Valongo trilobites could attain started to filter through the scientific community.

"As in a typical working quarry, the rock at the Canelas quarry is removed in layers," Couce said. "After all, this has long been a commercial operation, not one specifically designed to reveal fossils. The trilobite finds that have occurred at Arouca have been a side benefit of the process, not the primary mission of the workers. But everyone involved quickly realized what they had uncovered; the right authorities were contacted, and special care was taken to extract these specimens in the best possible manner."

The somewhat surprising fact is that a vast preponderance of Canelas quarry layers have proven to be virtually barren of fossils, possibly indicating that for much of the Middle Ordovician this was an oxygen-deprived marine environment. For many years, it was assumed that few, if any, significant paleontological finds would be made within quarry parameters. Indeed, it wasn't until after more than a century of commercial digging at the site that the layers featuring the Valongo Formation's oversized trilobites were uncovered . . . and almost immediately exhausted.

Only a few years after that special strata had been opened in 2009, workers ran into 35-meter-thick hillside overburden that basically stopped any fossil seeking efforts dead in their trilobite tracks. If traditional tile work is allowed to continue at the Arouca location, and rock is removed at the same slow and steady pace as it has been for much of the last 100-plus years, it is now estimated that it may take another three decades or longer before the trilobite-bearing layers are once again exposed and ushered into prominent play by quarry workers. However, digging of all sorts, whether for paving tiles or fossil treasures, may soon be brought to a forced conclusion at the Canelas quarry. At best, with the area now falling under UNESCO Geopark jurisdiction, removing any trilobite material from these Valongo Formation layers will become greatly reduced. At worst, such digging may be stopped completely. Either way, subsequent quarry discoveries face the possibility of being immediately labeled as "important national artifacts" and controlled by Portuguese authorities.

Those lingering legal limitations seem to have done little to douse local enthusiasm directed toward the Canelas locale and its grand fossil fauna. To further herald the quarry's recently acquired global notoriety, as well as to funnel additional tourist interest (and money) toward Arouca, the town has created an impressive on-site fossil museum. This facility is designed to display some of the outstanding specimens found over the last few years in the area's Ordovician layers, including large matrix blocks featuring congregations of possibly mating, possibly molting, giant trilobites. With all future digging at the Arouca outcrop now jeopardized by UNESCO's increased involvement, a unique paleontological conundrum has emerged: just as worldwide interest in these amazing Portuguese trilobites has begun to reach its apex, those unique fossils are at risk of again becoming something of an "endangered species."

Despite a spate of currently unresolved issues, enough of these special trilobites have already been recovered from the Canelas quarry's mudstone layers to keep academics and collectors busy for many years to come. These oversized Paleozoic prizes now rank among the most intriguing invertebrate fossils on Earth, specimens worthy of study and display in any museum from New York to New Delhi. These magnificent relics drawn from the rich hillsides of Portugal now enjoy well-deserved recognition as being among the true giants of the trilobite world.

RAPID REPORTS

Ontario, Canada

The various 460-million-year-old Ordovician-age formations that have emerged throughout Ontario, Canada—including the Lindsay, Bobcaygeon, and Verulam—rank high on any compendium featuring the world's leading trilobite sites. During more than a century of exploration and excavation by both professionals from Toronto's nearby Royal Ontario Museum and dedicated legions of local amateur enthusiasts, these varied fossiliferous outcrops have provided the trilobite community with some of the most impressive, fascinating, and scientifically significant examples of all time. Such notable species as *Gabriceraurus dentatus*, *Isotelus gigas*, *Ectinaspis homanolotoides*, and *Hemiarges paulianus* hail from the diverse yet geologically associated formations of the region and are usually three-dimensionally preserved with a dark brown calcite shell. In recent years, many of the vicinity's quarries have been closed due to safety concerns, making Ontario's already available trilobite material even more precious to those who truly appreciate the special morphological qualities exhibited by these amazing arthropods.

Beecher's Trilobite Bed, New York

For more than 120 years, *Triarthrus eatoni* trilobites of unique import and beauty have been found within the thinly striated, Upper Ordovician layers of the Lorraine Shale that occur in Oneida County, New York. What makes these small trilobites (usually no more than 3 centimeters in length) particularly remarkable is their dramatically pyritized preservation and the fact that many of these 450-million-year-old fossils present exceptional soft-tissue detail. Ever since Charles Emerson

ISOTELUS GIGAS (DEKAY, 1824)
Upper Ordovician; Richmond Formation, Arnheim Member; Mount Orab, Brown County, Ohio, United States; 12.3 cm

Compared to the rather prolific *I. maximus*, this is by far the rarer *Isotelus* species found at the legendary Mount Orab locale.

PARVILICHAS (?) N SP.
Upper Ordovician, Maysvillian Regional Stage; Lorraine Shale; Martin Quarry, Beecher's Trilobite Bed; Oneida County, New York, United States; 2.9 cm

This is a unique complete lichid found amid the traditional *Triarthrus* specimens in a privately owned quarry located adjacent to the famed Beecher's Bed.

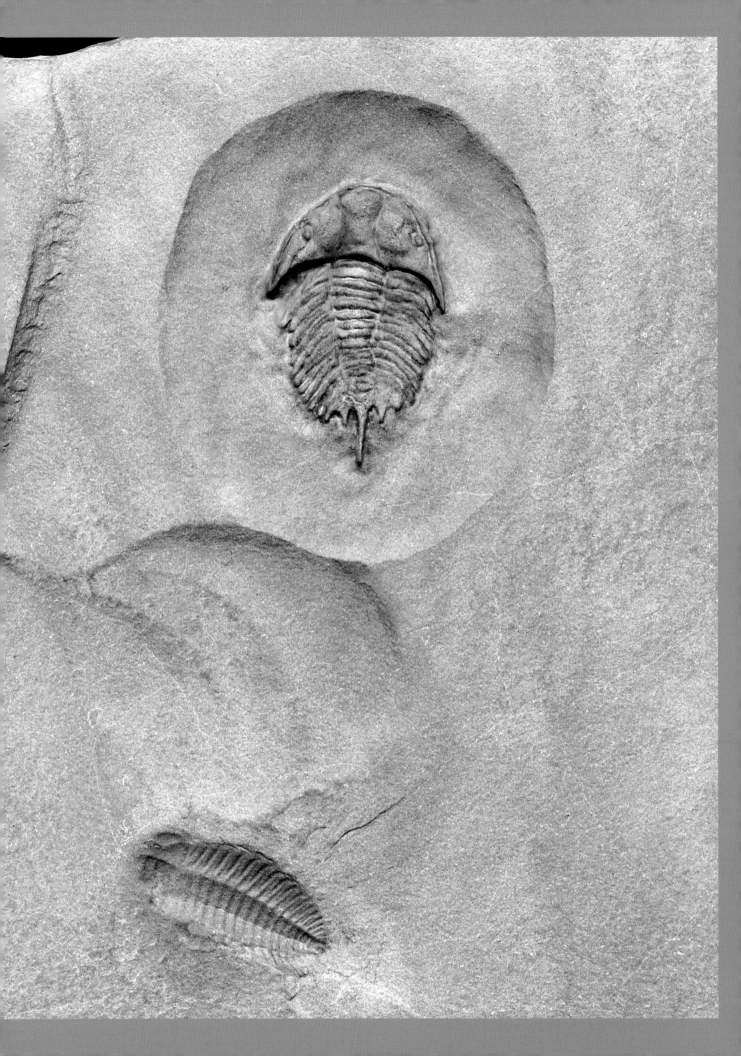

Beecher of Yale University began exploring this site during the last decade of the nineteenth century, these trilobites have been renowned for displaying the finest examples of trilobite ventral morphology ever found. Indeed, when carefully prepared, many specimens provide evidence of legs, antennae, gills, and even eggs. It was long believed that Beecher had exhausted the location during his three-year dig, and upon his death in 1904, it was assumed that the site, commonly known as Beecher's Trilobite Bed, had been lost forever. However, in 1984 amateur paleontologists Tom Whiteley and Dan Cooper, notably followed by researcher/preparator Markus Martin, rediscovered the location, and ongoing excavations have since revealed hundreds of additional complete *Triarthrus* specimens, along with, in rare instances, unidentified odontopleurids and lichids.

Mount Orab, Ohio

It's not often that a trilobite receives recognition as an official state fossil. As might be expected, dinosaurs enjoy such a designation in a number of regions across the United States, as do the occasional woolly mammoth and saber-toothed cat. But in Ohio, the large *Isotelus maximus* specimens that for more than a century have been pulled from the 447-million-year-old Ordovician layers that sit atop Mount Orab have earned that exact distinction. However, these impressive Arnheim Formation trilobites, which on occasion reach up to 35 centimeters in length, are not the only renowned species that have emerged from these rocks. *Flexicalymene retrorsa* ranks as the formation's most prolific trilobite, and the beautiful and rare *Allolichas halli* represents one of the true trilo-trophies

of the American fossil landscape. The formation also features an impressive assortment of crinoids, cephalopods, and brachiopods that lurk amid the relatively soft gray limestone layers. It is the giant ovate *Isotelus*, however, with its smooth chocolate-brown calcite carapace, that continues to draw trilobite enthusiasts back to Mount Orab year after year in hopes of finding the "ultimate" complete specimen.

Lady Burn, Scotland

Lady Burn has long been recognized as one of Europe's most famed and diverse Ordovician fossil localities. This site is adjacent to the small Scottish town of Girvan and lies along that country's rugged western coast, about 50 kilometers south of Glasgow. It is a rustic landscape filled with rolling green hills and rock-strewn outcrops that borders on the frigid waters of the colorfully named Firth of Clyde. For centuries, area residents have walked through this terrain while keeping a sharp eye peeled for the unique "fossil pockets" that characterize this legendary locale's 437-million-year-old sedimentary deposits. These finely grained pockets of hard, calcareous sandstone can yield well-preserved examples of starfish, crinoids, and bryozoans, in addition to more than a score of trilobite species. Lady Burn fossils are often coated in a sturdy brown or orange mineralized patina that serves to instantly distinguish the site's array of unusual trilobite species. These bugs include the likes of *Toxochasmops bissetti*, *Paracybeloides girvanensis*, and the giant cheirurid *Hadromeros keisleyensis*, which on rare occasions have been found as complete specimens up to 25 centimeters in length.

ORDOVICIAN TRILOBITE
PHOTO GALLERY

(ABOVE) *AUTOLOXOLICHAS THREAVENSIS* PHLEGER, 1936

Upper Ordovician; Ashgillian Stage, Rawtheyan Substage; South Threave Formation, Farden Member, 'Starfish Bed A'; Threave Glen, Scotland, UK; 6.3 cm

This is an unusual and particularly well-preserved specimen from western Scotland—one of the most prolific and famed Ordovician locales in all of Europe.

(OPPOSITE PAGE) *CERAURINUS MARGINATUS* (BARTON, 1913)

Upper Ordovician; Lindsay (Cobourg) Formation; Colborne, Ontario, Canada; 9.2 cm

This is an exceptionally well-preserved specimen of a legendary Canadian species. Late in the twentieth century these trilobites were relatively common components of any comprehensive collection. Since then they have become increasingly scarce.

ACIDASPIS CINCINNATIENSIS (MEEK, 1873)

Ordovician; Kope Formation; Pendelton County, Kentucky, United States; 5.1 cm

Among the most elegant trilobite species to emerge from the American Midwest, this is an exceptionally large example of a generally diminutive trilo-type. Preparing a spinose display like this requires up to 20 hours of work.

URALICHAS HISPANICUS TARDUS (VELA AND CORBACHO, 2009)

Late Ordovician; Izegguirene Formation; Bou Nemrou (El Kaid Errami); Morocco; 22.4 cm

This is an impressive trilobite that could reach 40 centimeters in length. Examples of this genera have also been found in Spain, Portugal, and France.

(ABOVE) *METOPOLICHAS PATRIARCHUS* (WYATT-EDGELL, 1866)

Lower Ordovician, Llanvirnian/Llandeilian boundary, Fairfach Group; Dinefwr Limestone, Murchisoni Biozone; Dinefwr Park Quarry near Llandeilo; Dyfed, Wales, UK; 5.4 cm

This area's sedimentary layers have long been extracted for use as building materials.

(OPPOSITE PAGE) *PARACYBELOIDES GIRVANENSIS* (REED, 1906), WITH ONE CALCICHORDATE (*SCOTIAECYSTIS CURVATA*); TWO BELLEROPHONTID GASTROPODS (*SINUITES SUBRECTANGULARIS*); ONE CEPHALOPOD (*CYRTOLITES SP.*); AND ONE ORTHID BRACHIOPOD (*LEPTAENA THRAIVENSIS*)

Upper Ordovician; Ashgillian Stage, Rawtheyan Substage; Upper Drummuck Group, South Threave Formation, Farden Member; "Starfish Bed," Threave Glen; Girvan, Ayrshire, Scotland, UK; 6.2 cm

Here is a nicely preserved example of *Paracybeloides* from the famed Girvan site in Scotland.

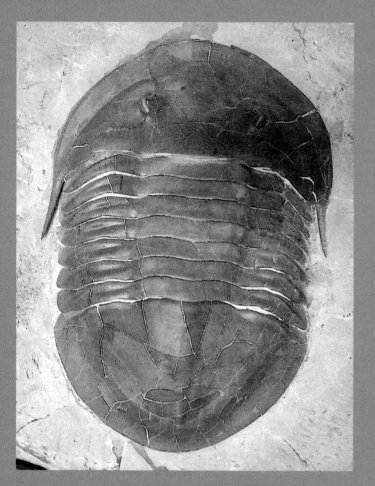

(ABOVE, LEFT) *ISOTELUS BRACHYCEPHALUS* **(FOERSTE, 1919)**

Upper Ordovician; Dillsboro Formation, Arnheim Member; Oldenberg, Indiana, United States; 22.8 cm

Here is a well-preserved example of this large asaphid, which closely resembles similarly sized *Isotelus maximus* specimens found in nearby Ohio.

(ABOVE, RIGHT) *XYLABION SP.*

Middle Ordovician; Sugar River Formation; Watertown, New York, United States; 5.3 cm

Found both in Ontario and New York in sites less than 200 kilometers apart, this is still not a scientifically recognized genus. Indeed, some academics believe it is a type of *Ceraurinella*.

(OPPOSITE PAGE) *BUFOCERAURUS BISPINOSUS* **(RAYMOND AND BARTON, 1913)**

Upper Ordovician; Verulam Formation; Kirkfield, Ontario, Canada; 5.2 cm

This represents one of the more unusual trilobite species to be found within the fossil-rich outcrops of eastern Canada.

(ABOVE) **ECCOPTOCHILE ALMADENSIS (ROMANO, 1980)**

Upper Ordovician; Llandeilo Series; Valongo, Portugal; 13.4 cm

Here is a large, colorful example of a rare European species. Note the unusual "break" in the middle of this specimen's thorax, indicative of a molted carapace.

(OPPOSITE PAGE) **GABRICERAURUS MIFFLINENSIS (DEMOTT, 1987)**

Middle Ordovician; Platteville Formation; Grant County, Wisconsin, United States; 7.3 cm

This exotic trilobite is often found in "pockets" that have yielded numerous complete examples. A major excavation conducted during the summer of 2014 uncovered nearly a dozen perfect specimens.

(ABOVE, LEFT) **PARACERAURUS INGRICUS (SCHMIDT, 1881)**

Middle Ordovician, Lower Llanvirnian; Kunda Regional Stage; Obuchov Formation; Putilovo Quarry; St. Petersburg region, Russia; 7.3 cm

This unusual species comes from a recently opened quarry where the coloration of both the trilobites and the surrounding matrix are notably different from "traditional" Russian examples.

(ABOVE, RIGHT) **ASAPHOPSOIDES YONGSHUNENSIS (LIU, 1982)**

Lower Ordovician, Tremadocian; Yinchufu Series; Yongshun, Hunan Province, China; 16 cm

In the later years of the twentieth century, this was among the first oversized trilobite species to emerge on the global stage.

(OPPOSITE PAGE) **CERAURINUS ICARUS (BILLINGS, 1860)**

Upper Ordovician; Whitewater Formation; Preble County, Ohio, United States; 5.7 cm

This is one of the rarest trilobites hailing from the famed Cincinnatian Lagerstatte of the American Midwest. It is unusual to find a specimen so perfectly preserved.

MEGISTASPIDELLA TRIANGULARIS (F. SCHMIDT, 1906)

Middle Ordovician, Upper Arenig Series; Huk Formation, Asaphus expansus Zone; Slemmestad, Norway; 11.2 cm

A species known from similarly aged outcrops in Russia, this is a rare complete specimen from a Norwegian site some 1,000 kilometers to the west.

(ABOVE, TOP LEFT) ***LEVICERAURUS MAMMILLOIDES* (HESSIN, 1988); TWO *CERAURUS GLOBULOBATUS* (BRADLEY, 1930)**

Upper Ordovician; Lindsay (Cobourg) Formation; Colbourne, Ontario, Canada; *(right)* Levi: 6.1 cm

This is a particularly well-preserved triple featuring two graceful cheirurid species.

(ABOVE, BOTTOM LEFT) ***SELENOPELTIS BUCHII* (BARRANDE, 1846)**

Upper Ordovician, Caradocian; Cabeço de Peao Formation; Maçao, Portugal; 3.2 cm

Found in a wide range of locations, this appealing positive/negative example is similar to species hailing from outcrops in England, Morocco, and the Czech Republic.

(ABOVE, RIGHT) ***ESTONIOPS EXILIS* (EICHWALD, 1858)**

Upper Ordovician; Kukruse Regional Stage; Viivikonna Formation, Kivioli Member; Narva, Northeast Estonia; 2.2 cm

This region's recently uncovered layers have yielded a smattering of dramatically preserved trilobites. The Upper Ordovician deposits are just now beginning to be thoroughly explored.

(ABOVE, LEFT) *LONCHODOMAS ROSTRATUS* **(SARS, 1835)**

Upper Ordovician; Kukruse Regional Stage; Viivikonna Formation; Narva, Northeast Estonia; 6.9 cm

This "needle-nosed" trilobite was captured in time alongside a neighboring bryozoan colony.

(OPPOSITE PAGE) *PSEUDOMERA CF. BARRANDEI* **(BILLINGS, 1865)**

Middle Ordovician; Antelope Valley Limestone; Nye County, Nevada, United States; 9.2 cm

This is one of the most distinctive trilobites found in the desert-dry environs of the American West. This is an exceptionally large example of a species that generally runs 4 to 6 centimeters in length.

(OPPOSITE PAGE, TOP LEFT) *HEMIARGES PAULIANUS* (CLARKE, 1894)

Ordovician; Verulam Formation; Ontario, Canada; 3.5 cm

This unusual species emerged from the rich fossil deposits that dot southern Ontario. For collectors, it is one of the most highly prized trilobites to be found within these sedimentary layers.

(OPPOSITE PAGE, TOP RIGHT) *CYBELELLA REX* (NIESZKOWSKI, 1857)

Upper Ordovician, Caradoc Series; Lowest Kukruse Regional Stage; Kingisepp Quarry; St. Petersburg region, Russia; 3 cm

It's only been during the first decades of the twenty-first century that extensive exploration of Russia's Upper Ordovician quarries has revealed beautifully delicate specimens such as this one.

(OPPOSITE PAGE, BOTTOM) *GABRICERAURUS SP.*

Upper Ordovician; Verulam Formation; Hastings County, Ontario, Canada; 9.2 cm

This is a strange trilobite that displays affinities to both *Gabriceraurus* and *Bufoceraurus*. Unfortunately, the shell has not been well preserved, so this specimen generates as many questions as answers.

(ABOVE) *CERAURUS N.SP.*

Ordovician; Lebanon Limestone; Belfast, Tennessee, United States; larger trilobite: 6 cm

This is an intriguing "double" of a species that is still undergoing scientific study and debate. Complete examples are rare from this location.

(ABOVE, LEFT) *KLOUCEKIA CF. ROBERTSI* (REED, 1904); *PSEUDOSPHAEREXOCHUS CONFORMIS* (ANGELIN, 1854)

Upper Ordovican, mid-Ashgillian; Lower Cautleyan Substage, Ddolhir Formation; Cynwyd Forest, Near Corwen, Denbighshire, Wales, UK; larger trilobite: 7.1 cm; smaller trilobite: 3 cm

This example is particularly noteworthy in that it presents a rare natural pairing of two relatively rare British trilobites.

(ABOVE, RIGHT) *LACHNOSTOMA LATUCELSUM* ROSS, 1951

Middle Ordovician; Upper Nine Mile Shale; Eureka County, Nevada, United States; 5.4 cm

Note the unusual eye placement on this medium-sized species, which makes it appear slightly cross-eyed. Examples of this trilobite have exhibited a surprising size range, from 2 to 10 centimeters.

(OPPOSITE PAGE) *GABRICERAURUS DENTATUS* (RAYMOND AND BARTON, 1913)

Upper Ordovician; Sugar River Formation; Lewis County, New York, United States; 8.2 cm

A species better known from Canada, this New York example is similarly impressive in size and shape.

THE SILURIAN PERIOD 3

444–419 Million Years Ago

ROCHESTER SHALE, MIDDLEPORT, NEW YORK:
UNLOCKING SILURIAN SECRETS

T IS A BARREN, 13-ACRE tract of land located near Middleport, New York—a sturdy stone's throw from Buffalo—a quarry that possesses all the inherent charm of a near lifeless landscape on some distant alien moon. The color scheme of this Rochester Shale outcrop is gray, upon gray, upon gray, with a hearty pale green weed only occasionally poking up through the solid rock to break the monochromatic monotony. Water has gathered in spots where the smooth, Silurian-age strata has been worn, or torn, away. During the all too brief Middleport summer (in these northern climes there are rumored to be only two seasons, the fourth of July and winter), tadpoles scurry about in small algae-covered pools, hoping to capture one of the countless insects that make this unforgiving, gravel-strewn repository their home.

Despite the initially forsaken appearance this property projects, the Middleport quarry is sacred ground. Wherever we walk on its hard but surprisingly fragile mudstone deposits, we unknowingly tread on ghosts from Earth's long forgotten past. Beneath 6 meters of stratified overburden—sedimentary rock that has accumulated since this area of

DICRANOPELTIS NEREUS (HALL, 1863)

Lower Silurian; Rochester Shale Formation; Caleb's Quarry; Middleport, New York, United States; 7.1 cm

Here is perhaps the most coveted trilobite found within the fossil-rich environs of the Rochester Shale. No more than a dozen complete examples have been uncovered during nearly 30 years of continuous digging at the quarry site.

Photo courtesy of the M. Shugar Collection

BUMASTUS IOXUS (HALL, 1852)

Lower Silurian; Rochester Shale Formation; Caleb's Quarry;
Middleport, New York, United States; 3.3 cm

This is a classic species originally described by the legendary
James Hall. Similar examples of this smooth-shelled trilobite
have been found everywhere from England to the Czech
Republic.

the world was part of a shallow ocean system 430 million years ago—the quarry is home to many of the most beautiful Silurian-age fossils in the world.

Silurian deposits are relatively rare in North America. The short (at least in geologic time) 25-million-year period stretching from 444 to 419 million years ago ranks behind longer and more revered Paleozoic intervals (the Cambrian, Ordovician, and Devonian) in terms of both its scientific renown and its availability to most fossil collectors. Yet in Middleport and much of the surrounding vicinity, the Silurian rules! From the legendary eurypterid quarry 100 kilometers to the

east to the famous shale deposits along the nearby Erie Canal, wherever a sharp-eyed fossil enthusiast may choose to look, the Silurian—a time of giant invertebrate marine predators and tiny, bizarre sea organisms—lives again.

Today the Middleport outcrop is the center of attention for a team of commercial fossil collectors, many of whom have been operating within the quarry's spacious confines for the better part of three decades. In their ongoing attempts to pry the area's Paleozoic treasures from their ancient homes, these intrepid souls often busy themselves for up to 10 hours a day, seven days a week, especially during the height of the region's limited digging season. Such dedication is easily explained: contained within the 3-meter-thick fossil-bearing layers of what has become known as Caleb's Quarry are perfectly preserved cystoids, magnificently detailed crinoids, and some of the most complete starfish to be found in North America—all just waiting to be freed from the rock-hard sediments that have encased them for eons.

Those primal relics only serve as a collective appetizer for the diggers' true quest. Their more focused efforts are directed toward one goal and are inspired by a solitary dream: the chance to recover some of the most celebrated and beautiful trilobites ever to crawl through Earth's early seas. Just mentioning *Arctinurus boltoni*, *Trimerus delphinocephalus*, *Bumastus ioxus*, and *Dalmanites limulurus* is often enough to satiate the arthropod-centric cravings of many savvy fossil fanatics. But for the Rochester Shale quarry's heavily calloused crew, mere names mean little—nothing

CALYMENE NIAGARENSIS (HALL, 1852)

Lower Silurian; Rochester Shale Formation; Caleb's Quarry;
Middleport, New York, United States; largest trilobite: 4 cm

This is an appealing and singularly distinctive grouping of a
relatively common species, perhaps the most abundant trilobite
found within the Caleb's Quarry locality.

ARCTINURUS BOLTONI (BIGSBY, 1825); DIMEROCRINITES SP (CRINOID)

Lower Silurian; Rochester Shale Formation; Caleb's Quarry; Middleport, New York, United States; 12.7 cm

Among the most storied of all American trilobites, complete examples of these large lichids were extremely rare until a commercial dig was begun in the 1980s.

short of the "real thing" will suffice. These admirably persistent middle-aged prospectors—most of whom either live nearby or make weekly round trips of up to 400 miles to pursue their paleontological passion—spend days on end repeatedly performing their laborious, blister-generating, back-bending, rock-breaking task, hoping to be the first human eyes ever to glimpse a certain fossil specimen.

Forget for a moment that large *Arctinurus* have long ranked among the most coveted trilobites on Earth, with relatively pristine examples often commanding between $1,000 and $3,000 on the open market, or that the rich Middleport layers are yielding once rare *Dalmanites* by the hundreds. These determined diggers continue their efforts for one primary reason—a pure love of what they do.

"I may be a bit biased, but to me this is the greatest job in the world," said Ray Meyer, who owns a lengthy lease on the Middleport quarry. "Each morning I get up ready to go, never knowing exactly what I might find out there. The thrill of the unknown is just part of the excitement of working the Rochester Shale. We've now been doing this for 30 years, but there's still a chance we'll find something we've never seen before."

People have been splitting rocks in and around the Middleport area for more than 200 years. Construction began on the Erie Canal in the early 1800s, a waterway that stretches for nearly 600 kilometers between Albany and Buffalo and connects the Hudson River with Lake Erie. Almost immediately canal workers operating near the dig's western terminus began discovering strange things within the Silurian-age blocks they were extracting. Some objects were recognized as being brachiopods and corals, but others were unknown, most notably the trilobites. Curious canal workers decided to hold on to a few of the more unusual rocks they encountered until they had a chance to show them to their superiors. Unfortunately, many early specimens were quickly discarded by Erie Canal officials, few of whom wanted their underlings to be distracted from the important task at hand. Despite such corporately imposed barriers, a significant number of fossil finds were made during the Erie Canal dig, and a good number of examples uncovered during that landmark project still reside in the nearby Buffalo Museum of Science.

In the 1830s, one of America's most noted paleontologists, James Hall (later the director of the Museum of Natural History in Albany), began the first extensive scientific excavation of the fossil-rich Silurian outcrops of western New York State. Using the sedimentary exposures created by the canal dig, which was completed in 1825, Hall collected an amazing array of invertebrate fossils. His discovery of a nearly intact *Arctinurus* specimen in 1834 created quite a cultural uproar, triggering many community groups (including leading religious factions) to question exactly what this strange, somewhat leaf-shaped, ancient creature might actually be. So much pressure was placed on Hall and his benefactors, especially by politicians who looked askance at his findings, that his funds were frequently cut off by the State of New York. Thus, it wasn't until 1847, when he published the landmark three-volume treatise *The Invertebrate Fossils of New York*, that many of his historic discoveries finally came to light. With the appearance of that seminal work, the legend surrounding the Rochester Shale deposits of upstate New York began to take shape.

For the next 140 years, diggers and collectors from throughout the Northeast would frequently wander up to the Erie Canal cuts, wait for a canal lock to drain to the required level, and attempt to rapidly extricate a few fossil prizes from the slippery Silurian exposures before the lock again began to fill. This work was, at best, an iffy proposition, and recovery of a complete trilobite was certainly a rare occurrence. It is estimated that no more than a half dozen articulated Rochester Shale *Arctinurus* specimens had been found prior to the full-time Caleb's Quarry operation that began in 1991.

"There's a lot of history involved with the region and with these trilobites," Meyer stated. "When you dig in these deposits, you sometimes just need to stop and think about the past."

Working the striated sedimentary layers of the Middleport quarry is far from easy. Twenty-pound crowbars and hefty sledge hammers are the standard tools of this trade, and those that wield them often pay the price for their efforts in sore muscles and aching joints. Despite the Silurian strata's incredible longevity, once the platter-sized sheets of gray matrix are excavated—on occasion, by the quarry's lone, highly temperamental backhoe—they can quite literally turn into a fragment-filled powder if exposed to little more than a steady night's rain. But those who regularly move rock at this fabled site are more than willing to suffer through such minor inconveniences in exchange for the fossiliferous rewards their efforts can generate.

On any given summer's day, as many as eight eager participants may be simultaneously digging in different parts of the quarry. Some visitors perform scientifically inclined research via millimeter-by-millimeter excavations along the site's tree-lined perimeter. A few investigate the location's "west wall" for fossilized evidence of starfish and crinoids, and others rigorously mine the quarry floor for marketable examples of the area's most common trilobites, *Dalmanites limulurus* and *Calymene niagarensis*. Still others seek unsullied Silurian terrain in their quest to unearth rare species to add to their private collections. But those hunting for unusual trilobites within the quarry's borders are all too aware that finding a "keeper" is far from guaranteed. Meyer is willing to divulge that he has occasionally gone for as long as a week—extracting, examining, and tossing aside up to two tons of rock a day strictly by glove-covered hands—without encountering a single complete specimen.

Disarticulated *Bumastus* heads and *Trimerus* tails are relatively common finds. These trilobites, like all others, molted their calcified outer shells at regular intervals, leaving behind many tantalizingly fragmented fossilized remains. But good, undamaged examples of the up to 15-centimeter-long *Arctinurus*, 20-centimeter *Trimerus*, or four-centimeter *Bumastus* are indeed atypical. On one notable occasion, Meyer found four complete *Arctinurus* specimens in a single day, and a recent quarry visitor landed two perfectly

articulated *Trimerus* in a 30-minute stretch. However, these moments of Paleozoic glory are few and far between. It is the memories of such events, though, that inspire many resolute collectors out into the 90° Fahrenheit (32° Celsius) July heat or the subfreezing October cold (the quarry operates roughly from late April until early November) in search of their elusive Silurian treasures.

"Digging is something you either love or you hate," Meyer said. "There are some people who can't get enough of it, and there are others who can't stand it. But if you love fossils, I think as soon as you find something that's been hidden in these rocks since the Silurian, you'll be hooked."

Freeing these trilobites from their thickly layered sedimentary casings is only the first step in a complex process of properly preparing that fossil specimen for study, sale, or display. On more than one occasion, extremely rare examples (including a near perfect *Dicranopeltis nereus*, a 6-centimeter lichid that is perhaps the quarry's most coveted trilobite) have been destroyed by a careless hammer blow or a poor excavation technique. Electric saws and sturdy rock chisels are often utilized to remove specimens from the surrounding matrix, but even then totally clean, unblemished examples are the exception rather than the rule. Usually, even the most complete trilobites emerge as a jigsaw puzzle assembly of pieces, perhaps with eyes or parts of their outer shell still encased in the negative side of the fossil-bearing rock. These examples are in desperate need of the surprisingly delicate touch provided by a well-trained trilobite

BUMASTUS SP.

Lower Silurian; Rochester Shale Formation; Caleb's Quarry; Middleport, New York, United States; 11 cm

There has been recent debate concerning whether this large variety of *Bumastus* represents a different species than the common *B. ioxus* . . . or perhaps even a different genus.

"prepper" before they can regain at least a modicum of their former glory.

Preparing a bug from the Middleport quarry often requires 5 to 10 hours of detailed work, removing clinging bits of matrix, realigning disarticulated segments, and painstakingly reattaching tiny fragments of shell. But more often than not, the results are well worth the effort. When finished, the trilobite specimens possess a beautiful charcoal-gray exoskeleton and retain much of their original 3D appearance. Some even exhibit evidence of epibionts, small brachiopods that attached themselves to the trilobite's exoskeleton during its life along the ancient ocean floor and became fossilized right along with their host. Most Rochester Shale trilobite species are thought to have been bottom dwellers. Thus, they either used their intricate, widespread morphological forms (*Arctinurus, Dicranopeltis*) to settle on the seabed like a modern-day flounder or maneuvered their smoothly shaped bodies (*Bumastus, Trimerus*) to burrow into the surrounding seafloor silt to avoid menacing predators.

"I love preparing a trilobite—especially after I dug it up," said the late Eugene Thomas, who prior to his passing in 2009 was both the principle preparator for the material emerging from the Middleport quarry and a renowned weekend digger. "Sometimes you really don't know what you have until you start putting it all together. Finding a good specimen is really fun, but restoring it to top condition is a close second."

With the unparalleled quality and quantity of trilobite material being recovered from Middleport's Rochester Shale quarry, scientists and collectors from around the world are being provided with a unique glimpse of what life may have been like in those long-ago Silurian seas. Beautifully fossilized *Arctinurus* exoskeletons showcasing bite marks from a still unknown predator (some wounds displaying healed or partially regenerated features),

Dalmanites multiplates possibly exhibiting sexual dimorphism, and *Trimerus* growth series specimens ranging from less than 2 centimeters to over 18 centimeters in length have combined to present the paleontologically inclined with an unprecedented depth of information concerning this thriving mid-Paleozoic ecosystem.

Quite simply, the finds at the Middleport quarry are helping to bring "alive" animals that have been gone from the face of our world for more than 400 million years. It is exciting, unpredictable, and satisfying work, and the incredible trilobite specimens being unearthed more than counterbalance the difficult labor involved with each and every discovery. For those with a strong back and a healthy imagination, the notion of recovering these unique treasures from our planet's distant past holds a fascination few other earthly delights can equal.

TRILOBITE FEEDING BEHAVIOR

It's not a great surprise that the subject of trilobites rarely arises in conversation outside museums' inner sanctums and fossil show backrooms. When it does, however, especially among people not particularly familiar with the fundamentals of these ancient marine organisms, two rather unusual questions tend to predominate. One is what trilobites tasted like . . . the other is what they ate.

Since trilobites have been gone from Earth for roughly a quarter of a billion years, it seems safe to surmise that we'll never know the answer to the first of these queries, although many trilobite hobbyists admit to clandestinely hoping that they tasted a bit like their arthropod cousin the lobster. As to wondering what trilobites may have eaten in those long-gone Paleozoic seas, if we rely on a bit of basic biological logic combined with a smattering of empirical evidence, we can at least begin to present an educated guess.

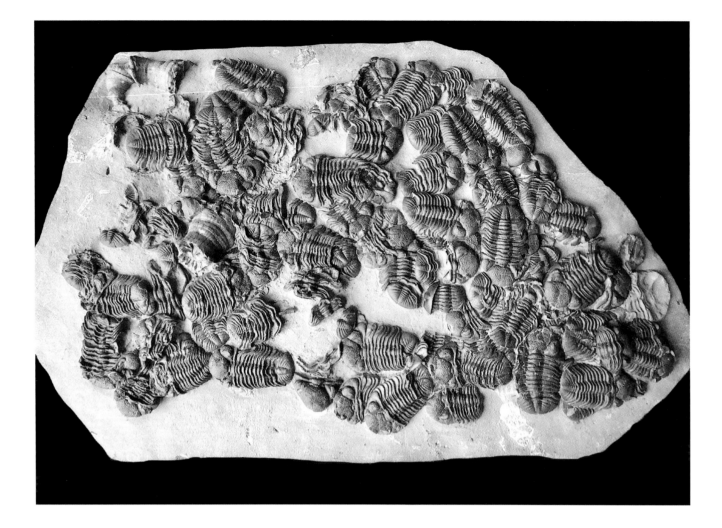

We know that trilobites occupied a wide variety of marine habitats during their lengthy passage through time—ranging from shallow tropical reefs, to cool continental shelves, to murky light-deprived depths. Thus, their available food resources naturally changed within each of those aquatic environments. It is currently believed that some trilobites were filter feeders, feasting on free-floating swarms of proto-plankton, and others may have used their hypostome (mouth plate) to scrape parasites from undersea rock surfaces. Some may have survived as scavengers, ingesting the remains of primitive worms, ancestral jellyfish, or whatever else may have fallen into their deep ocean lairs.

A good number of trilobites, however, were most likely predators, with some perhaps even being cannibalistic, feeding on a diverse variety of smaller trilobite species. Especially early in their history, larger trilobites such as *Glossopleura gigantea* and *Olenoides superbus*—with

ELDREDGEOPS RANA (GREEN, 1832)

Middle Devonian (Givetian); Hamilton Group, Moscow Formation, Windom Shale Member; Penn Dixie Quarry; Hamburg, New York, United States; average size: 3 cm

This is an impressive assemblage featuring one of the more common trilobite species found in the eastern United States. It is speculated that *Eldredgeops* often gathered for feeding, mating, or molting.

individuals attaining lengths up to 15 centimeters—may have preyed upon more diminutive trilo-types. Indeed, fossilized examples of common species such as *Elrathia kingii* and *Asaphiscus wheeleri* often bear the apparent scars of such encounters on their carapaces. Trilobites also may have fed on the multitudes of soft-bodied creatures that often surrounded them in the Cambrian and Ordovician seas. Although no direct evidence of such aggressive trilobite behavior has yet been revealed in the fossil record, it seems reasonable to assume that such a predator-prey relationship would exist between creatures living in such close proximity.

On a somewhat tangential subject, it has recently been postulated in certain academic circles that the net biological by-product of all this feeding activity (colloquially known as "trilobite poop") may have played a vital role in the development of other life-forms within those primal seas. Some scientists believe that the sheer amount of excrement produced by the legions of arthropods inhabiting the early Paleozoic oceans generated enough fertilizer to provide primitive plant life with the needed impetus to develop and expand its realm. The key result of this vegetative expansion was a significant increase in the oxygen levels of the Cambrian seas. This oxygen-generating activity slowly enabled the Earth's environment to become more hospitable to a wider variety of organic forms—both in the oceans and eventually on dry land.

It can even be argued that without the early success of the trilobite line, and the initial development of their digestive tracts, life on our world would be far different than it is today. If it wasn't for primitive trilobite feeding habits, along with their resultant pellets of excrement, Earth's early faunal and floral experiments may have ended in a far more abrupt and dire fashion. How fortunate for all subsequent life-forms, including humans, that they did not.

WREN'S NEST, DUDLEY, ENGLAND: *CROWN JEWEL OF THE TRILOBITE KINGDOM*

The scene would have appeared somewhat strange to our contemporary eyes. There they all were—men, women, and children—attired in their mid-nineteenth-century Sunday's Finest, scurrying through the rocky, Silurian-age outcrops of Wren's Nest, located near the quiet town of Dudley. They had all come to this hilly stretch deep in the heart of England's coal-rich Black Country in search of one thing—what seemed to be insects that had been miraculously turned to stone.

Some among this intrepid group of incongruously stylish visitors may have had the foresight to carry along a small hammer with which to gently break apart any interesting rocks they might encounter. Others may have brought pocket-sized magnifying lenses through which to better view their potential finds. Most, however, contentedly strolled about the rugged terrain of Wren's Nest unencumbered by the traditional tools of collection. They were there strictly for the sheer joy of adventure and discovery, hoping that a prized trilobite specimen would both literally and figuratively fall under their well-shod feet.

Today we might find the notion of prospecting for fossils without the requisite array of work boots, rock chisels, protective glasses, and padded knee guards somewhat disconcerting, if not downright dangerous. But in the middle years of the 1800s, when such high society outings were a weekly occurrence at these now famed Much Wenlock Limestone deposits, hunts for what were then known as Dudley Locusts—today recognized as the trilobite *Calymene blumenbachii*—were on the cutting edge of both family holiday fun and true scientific research.

Quite a few of those early adventurers grew so enamored with the incredibly diverse and highly

BALIZOMA CF. VARIOLARIS (BRONGNIART, 1822)

Middle Silurian, Wenlockian Stage, Homerian Substage; Much Wenlock Limestone Formation; Wren's Nest Hill, Dudley, West Midlands, England, UK; 4.5 cm

The stalked eyes adorning this species—with one being damaged—allowed it to cover itself with sediment while still observing the surrounding seafloor, a distinct evolutionary advantage.

CALYMENE BLUMENBACHII (DESMAREST, 1817)

Middle Silurian, Upper Wenlockian Stage; Much Wenlock Limestone Formation; Wren's Nest Hill, Dudley, West Midlands, England, UK; 5.5 cm

This is the legendary "Dudley Locust" of the English Midlands. For many years, a figured representation of this species appeared on the Dudley town crest.

detailed fossil fauna to be found at Wren's Nest that they began to spend more and more of their free time scouring the location for Paleozoic bounty. In the process, these fortunate and resourceful individuals managed to amass some of the most impressive Dudley collections ever recorded. In fact, such was the era's enthusiasm for the Wren's Nest site that when the noted Scottish geologist Sir Roderick Murchison (author of the groundbreaking 1839 volume, *The Silurian System*) would lecture on Dudley's fossils, his talks often attracted incredibly large throngs of enthusiastic listeners. One such gathering, held at Wren's Nest itself, drew an estimated crowd of 15,000 fossil frenzied folks and generated national headlines in the British press.

Over the ensuing decades, hundreds, if not thousands, of complete trilobite specimens representing more than 80 species have been recovered from the Wren's Nest locale. Many of these beautifully preserved Silurian remnants have ended up as star attractions in private collections around the

globe, and others have been handed down from generation to generation as prized British family heirlooms. As one might expect, an amazing assemblage of trilobites drawn from the area's prolific Wenlock Limestone also resides in the cloistered cabinets of London's famed Natural History Museum.

"Many of the trilobites that were unearthed centuries ago have been maintained in private cabinets of curiosity to this very day," said Alf Cawthorn, a recognized authority on British trilobites. "It's still not unusual to see a Dudley fossil come up at an English estate sale or auction. A number of *Calymene blumenbachii* actually became the centerpieces for jewelry back in the nineteenth century."

Silurian-age sedimentary outcrops found in and around the Dudley region have been exhumed, examined, and collected for nearly a millennium. During that time, the Wren's Nest locality has continually grown in prominence and now ranks among the most renowned paleontological repositories in all of Europe and as one of the most prominent Silurian strata in the world.

Quarrying the area's thick limestone layers was not initially undertaken with fossils in mind. Rather, it was done to acquire building materials for some of the town's most ambitious architectural projects. Many of the vicinity's famous landmarks—including nearby Dudley Castle built early in the twelfth century—were constructed solely out of the 420-million-year-old limestone blocks brought forth from the nearby hillsides. Throughout the

CYBANTYX ANAGLYPTOS LANE AND THOMAS, 1978

Middle Silurian, Wenlockian Stage, Homerian Substage; Much Wenlock Limestone Formation; Wren's Nest Hill, Dudley, West Midlands, England, UK; 6.5 cm

This species represents one of the more unusual members of the Dudley fauna. Rarely did trilobites at this location grow to such impressive dimensions.

following centuries, these rocks have also proven invaluable for a wide range of other purposes—initially as a form of quicklime fertilizer for many area farms and later as an essential ingredient in manufacturing both iron and steel, for which the English Midlands are justly famous.

It is understandable that those who first explored these Silurian layers 900 years ago were both confused and confounded by the numerous fossils they encountered. They may have recognized a brachiopod, a tabulate coral, or even a flower-mimicking crinoid for what they were—or at least what they *thought* they were—but a trilobite? There is simply no way these early pioneers could have realized that they were having a close encounter with the signature life-form of Earth's early history, a creature that had lived during an age when this corner of Britain looked radically different than it does today. To them, these strange "insects made of stone" must have appeared to be bizarre talismans from some distant and mysterious time.

It is also highly unlikely that any of these early Dudley devotees—virtually all of whom believed the world was only a few thousand years old—possessed even the slightest clue regarding the true age and significance of these primeval relics. Even fewer would have dared to imagine that in the Silurian the vicinity that now comprises the English Midlands was situated beneath a shallow, semitropical sea—a body of water that would have totally enveloped the picturesque hillsides upon which they then stood. Yet their lack of such insight did little to deter the fossil-fueled enthusiasm displayed by these spirited adventurers.

They, along with others attracted to the Dudley environs by the abundance of unusual and often inexplicable discoveries made within these mid-Paleozoic rocks, continued to regularly search through the Wren's Nest exposures for the next 600 years. However, it wasn't until the dawning of the 1700s that a few daring and enterprising souls

DEIPHON BARRANDEI (WHITTARD, 1934)

Middle Silurian, Wenlockian Stage, Homerian Substage; Much Wenlock Limestone Formation; Wren's Nest Hill, Dudley, West Midlands, England, UK; 3.2 cm

A bubble-nosed cheirurid, this ranks among the stranger members of the renowned Dudley biota. Found early in the twentieth century, complete examples are among the "crown jewels" of any trilobite collection.

began to more closely investigate these outcroppings and appreciate the wide variety of fossilized life-forms—especially trilobites—that they held. According to reports made at the time, eons of weathering of the hard limestone layers had freed so many disarticulated trilobite fragments from their primeval burial sites that they practically littered the ground. Over the following centuries, as interest in the natural sciences began to slowly expand within England's upper classes, collecting fossils throughout the Wren's Nest vicinity became something of a cottage industry. By the mid-1800s, many local residents would scour the site for fossil material and set up small roadside stands that offered a variety of choice Paleozoic morsels as souvenirs to those well to do weekend visitors not inspired to search for such items themselves.

"There are stories of families leaving directly from church on Sunday afternoon, and taking a leisurely coach drive up to Wren's Nest," said former Dudley resident, Simon Taft. "There they would proceed to walk throughout the area searching for fossils and other oddities. Some locals would tend to this trade by selling their wares—including their own self-collected fossils—along the road to and from the main rock exposures."

The fact that these Dudley specimens are Silurian (named after an ancient British tribe that inhabited what is now Wales) makes them of particular significance to both collectors and academics. That 25-million-year-long age was the shortest, and perhaps the least "glamorous," of the Earth's various Paleozoic periods. Unlike the Cambrian, which holds special scientific cachet for bearing witness to the initial flowering of multicellular life, or even the Devonian, during which myriad familiar looking creatures began to rear their often bizarre heads, the Silurian was a period marked most notably by global climatic stabilization and a significant rise, then fall, in worldwide sea levels. Judging by the rich diversity of faunal forms congregated within Dudley's Much Wenlock outcrops, it is easy to surmise that the area's climate in the mid-Paleozoic was quite different from the blustery cool temperatures that dominate the region today.

Recent magnetic mappings of the vicinity show that present day Dudley lay within 30 degrees of the equator during the Silurian. Fossils of more than 600 distinct species, including corals, echinoderms, crinoids, and gastropods—as well as the ubiquitous trilobites—provide further proof that this was once a warm coastal plane, a reef-filled tidal sanctuary that provided a near perfect environment for the proliferation of creatures both common and exotic. During the Silurian, the seas covering what is now western England were brimming with life, and predatory armored placoderms and ancestral jawless fish routinely patrolled those waters. But perhaps more than anything else, the fossil record makes it clear that the tides then enveloping Dudley were a haven for trilobites, with species such as *Cybantyx anaglyptos*, *Dalmanites myops*, *Acaste inflata*, and *Encrinurus tuberculatus* thriving amid the rich ocean currents.

Centuries of excavations have taken place at the Wren's Nest site, and possibly no other paleontological pocket in the world—including such equally bountiful and oft-studied Silurian outposts as New York State's Rochester Shale, and Sweden's Hemse Formation—have had their fauna as well chronicled. Yet perhaps there is something even more significant to collectors than this documented diversity of Dudley trilobite species; that is the magnificent manner in which these ancient organisms were preserved once their carapaces sank into the lime-rich mud that then covered the bottom of the marine basin. After all, many sedimentary locales throughout the world are filled with fossils. Too often, however, these remnants are little more than disarticulated bits and poorly preserved pieces of the region's once indigenous life. Because of the apparently gentle nature of the ocean currents that once surrounded the Wren's Nest environs, Dudley's future fossils were afforded the luxury of resting in a relatively undisturbed state as they were slowly covered by layer upon layer of mineral laden sea silt.

Over the ensuing eons, these sediments were gradually transformed into well-defined sheets of limestone strata by the internal forces housed within the planet. Yet despite the incredible compression and shearing torque that so frequently accompanies fossilization, the trilobites of Dudley have become world renowned for their amazingly lifelike, three-dimensional preservation.

Their fossilized exoskeletons have been replicated in a beautiful chocolate-hued calcite, with their detailed compound eyes (some featuring over 100 individual lenses) and delicate spines captured in minute detail. And after some skilled preparation work is done to remove what remains of their surrounding hard-rock matrix, today these ancient sea creatures appear little the worse for their 400-million-year odyssey through time.

These days most major natural history museums and private collections feature an assortment of trilobite specimens derived from these fossil-filled Much Wenlock Limestone layers. The pervasive Dudley Locust is still the most famous, and most common, species found, but an almost overwhelming array of other trilo-types have been discovered within the Wren's Nest site. These species range from the large homalonotid *Trimerus delphinocephalus*, which has been known to exceed 12 centimeters in length, to the diminutive and exceedingly rare bubble-nosed cheirurid *Deiphon forbosi*, which rarely grew larger than 3 centimeters.

Somewhat surprising, many of the trilobite examples found in the nineteenth century—when the concept of preparation often involved the hand wielding of anything from small pocketknives and spikelike nails to rather cumbersome dental tools—still hold their own in terms of quality, rarity, and desirability. In fact, the trilobites unearthed during those early digs retain a special cachet among enthusiasts who often seek out these so-called Victorian Collection pieces and choose them over

more recently discovered air-abrasively prepared Dudley examples.

The Wren's Nest location has now been designated an English Heritage Site. Anything more than surface collecting is banned, and even that is often frowned upon by local authorities. Yet the singularity, beauty, and scientific significance of the area's abundant fossil fauna is now more appreciated than ever. Whether uncovered 200 years ago or last Thursday, Wren's Nest trilobites remain among the most admired and desired in the world, especially by Paleozoic connoisseurs who revel in both the history and the allure exhibited by these magnificent examples of nature's handiwork. So even two centuries after those well-dressed Sunday visitors began exploring the area's rich Silurian outcroppings, Dudley, England, proudly retains its title as one of the planet's premier paleontological sites, a distinction it will not relinquish any time soon.

SEXUAL DIMORPHISM IN TRILOBITES

A few years ago, while strolling through a local rock and mineral show with friends, one in our group stated with straight-faced sincerity that he was about to begin writing a thesis exploring "sexual dimorphism in trilobites." Before any of us could wipe the somewhat quizzical looks from our fossil-lovin' faces, he quickly added that the subtitle of his work was going to be "Prove Me Wrong . . . If You Can!"

Perhaps our friend was just pulling our proverbial leg. All we know is that none of us has yet to see his finished paper, and we're beginning to doubt we ever will. Despite countless hours of high-minded scientific research on trilobites, next to nothing is known about the sexual nature of these intriguing invertebrates. And, in all honesty, that situation is not likely to change any time soon.

ENCRINURUS TUBERCULATUS (BUCKLAND, 1836)
Middle Silurian, Upper Wenlockian Stage; Malvern, England, UK; 5.5 cm

This is a classic species of the Wenlockian. Examples can still be found amid well maintained, Victorian-age Cabinets of Curiosities.

Although such a titillating topic would appear to be ripe for academic study, zero work has so far been done on the subject, and with good reason. The available Paleozoic material, or lack thereof, would make any effort little more than a well-intentioned exercise in trilobite-fueled futility.

From the moment these primeval organisms first appeared on the Cambrian scene some 521 million years ago, paleontologists logically assumed that trilobites were divided into males and females—although some modern members of the arthropod phylum do reproduce asexually. Those academicians also surmised that the trilobite mass mortality plates that on occasion have been uncovered in locales ranging from Oklahoma, to Russia, to Morocco, to China might very well provide evidence of early mating assemblages. Some scientists have even made direct reference to the existence of trilobite egg sacks and mating claspers, and Ordovician-age trilobite eggs have recently been discovered among pyritized *Triarthrus eatoni* specimens found in upstate New York. Other than these rather meager informational tidbits, however, we are left devoid of material that might provide either insight or substance on the topic.

A few leading authorities, including the renowned British paleontologist Richard Fortey, have speculated that the fancy frills, quills, spines, and horns that adorn certain trilobite species may represent a primal example of peacock-like sexual adornment. With a little imagination, one can readily picture male trilobites scuttling across the ancient seafloor brandishing an array of eye-catching anatomical decorations (perhaps in a variety of bold colors) to gain the notice of available females. This is still highly speculative, however, and pushes us to examine an equally compelling question: If certain male trilobites exhibited such gaudy structural ornamentation, have we possibly misclassified, or at least misidentified, the less ornate females of the same species? Judging by the diversity of morphological material presented in the fossil record, such a conclusion seems to be at least a debatable possibility.

Although our knowledge regarding sexual dimorphism in trilobites is currently negligible, we may eventually garner additional knowledge on the subject. Each year more trilobites displaying soft-tissue preservation are discovered in various Paleozoic outposts around the globe. And with improved preparation techniques revealing previously unknown details of trilobite anatomy—including digestive tracts and gill alignments—perhaps one day we will uncover key evidence that will shed additional light on this most basic of trilo-topics.

The possibility exists, however, that we may never learn how to properly differentiate male and female trilobites, and the planet will probably keep spinning on its slightly tilted axis even if we don't solve this primeval mystery. When our thesis-writing friend asks the academic world to "prove him wrong," perhaps the best anyone can currently do is admit that a few trilobite mysteries seem destined to remain secrets of the Deep Time past.

ARCTINURUS BOLTONI (BIGSBY, 1825)

Lower Silurian; Rochester Shale Formation; Caleb's Quarry; Middleport, New York, United States; largest: 13.5 cm; smallest: 9.5 cm

This is a highly unusual assemblage of four medium to large trilobites. Some observers have speculated that such size differences may be indicative of sexual dimorphism. Others, however, scoff at the notion.

ANTICOSTI ISLAND, QUEBEC: A TRILOBITE TREASURE TROVE

Some people will go just about anywhere and do just about anything to get their hands on a trilobite. No further proof is needed than knowing that explorers have ventured to remote Anticosti

Island, located amid the dangerous waters that comprise the Gulf of St. Lawrence in Eastern Quebec, Canada, in search of these ancient arthropods for more than 160 years. Indeed, the raging, often ice-clogged river currents that surround the island are so treacherous that the entire vicinity has earned a nasty reputation as the Cemetery of the Gulf due to the more than 400 shipwrecks that have been reported there.

The spectacularly preserved trilobite material found within Anticosti's abundant 430-million-year-old Silurian layers continues to serve as a major lure for both scientists and amateur collectors. Yet even today, merely landing your feet squarely on this oft-frozen isle can prove to be a daunting challenge. The nearest major airport is 700 kilometers away in Quebec City, and a small airstrip on the island itself offers only a smattering of seasonal charters. Perhaps the one sure way to reach this wayward destination is to rent a sturdy SUV in the provincial capital and drive approximately 320 kilometers to the picturesque riverside town of Rimouski. From there you and your vehicle will be loaded onto a once-a-week ferry and taken on a rugged 24-hour, seasickness-inducing voyage over to Anticosti itself. Despite the ship's basic but comfortable amenities and rather reasonable $240 roundtrip fee, it is not a journey for the faint of heart or weak of stomach. A more isolated subarctic spot in eastern North America than Anticosti Island might be difficult to find, but what is braving a few logistical inconveniences and overcoming some daunting physical barriers compared to the prize at stake—access to some of the continent's richest trilobite faunas?

The raw natural beauty, the verdant boreal forests, and the diverse wildlife (featuring 132 bird species, including bald eagles) make this a paradise for adventurers, campers, and hikers as well as trilobite enthusiasts. There's even a 355-square-kilometer national park on the island where visitors can see, and even hunt, everything from moose to white tail deer, both of which were introduced to Anticosti late in the nineteenth century and—without the threat of natural predators—have flourished ever since.

Once you reach the island, you are faced with the perhaps even more formidable task of navigating a large body of land (220 kilometers long and 50 kilometers wide) with few paved roads and minimal electricity, but riddled with fossil-bearing formations. These sedimentary layers include a variety of equally abundant Ordovician outcrops in addition to the dominant Silurian rocks. Together, these formations are more than 2,000 meters thick in certain spots on Anticosti, marking them as the most complete geological strata of corresponding age in the entirety of North America.

With a year-round population of only 240 intrepid souls, many of whom are involved in the island's upkeep and management, it is safe to assume that trilobites far outnumber the humans on this imposing tract of terra firma. Of course, the trilobites on Anticosti Island have long since been turned into fossilized impressions on stone . . . but what fossils they are!

In the Silurian, the creatures inhabiting the seas covering Anticosti were part of a warm-water, offshore ecosystem brimming with biodiversity. And whether it was due to a minimal number of lurking predators, slowly changing environmental conditions, or some other unknown evolutionary twist, many of the trilobites living in this rich sanctuary

DIACALYMENE SCHUCHERTI (TWENHOFEL, 1928)

Early Silurian, Llandoverian (Late Aeronian—Early Telychian); Jupiter Formation, Cybèle Member; Anticosti Island, Quebec, Canada; 9.2 cm

These large, impressive calymenid trilobites represent a classic example of Anticosti megafauna. On average, they are three times the size of similar species found throughout North America.

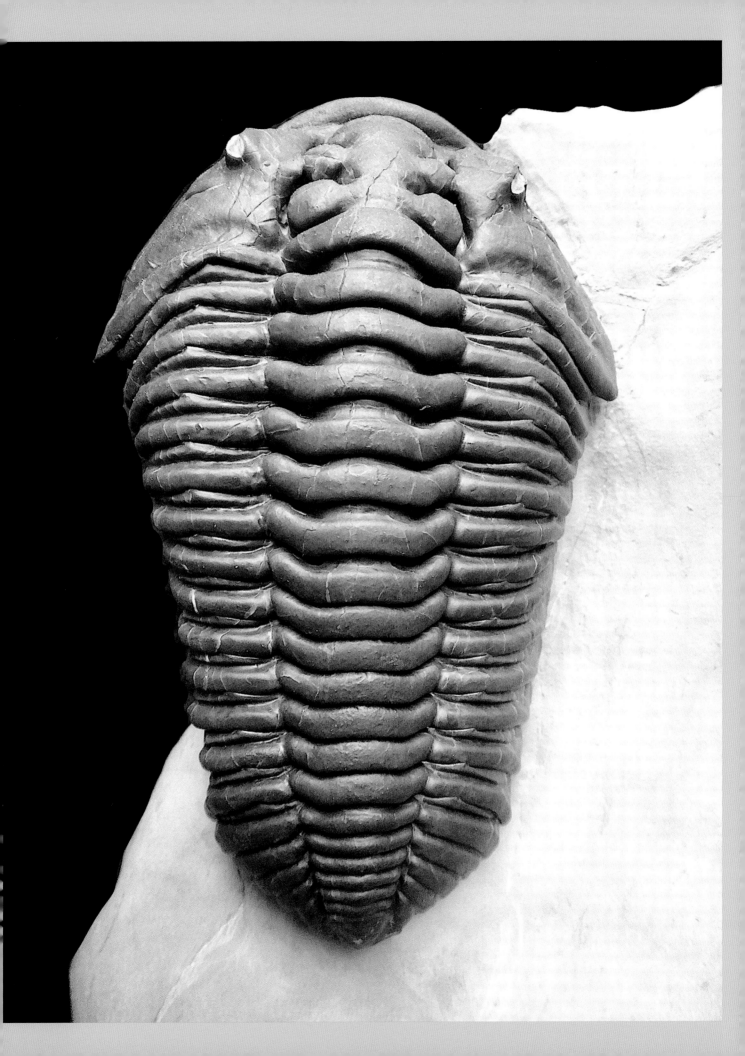

ECTINASPIS SP.

Early Silurian, Llandoverian (Late Aeronian—Early Telychian);
Jupiter Formation, Cybèle Member; Anticosti Island, Quebec,
Canada; 4.3 cm

This specimen was discovered in the 1980s lurking amid cliffside debris—a common Anticosti collecting method. Species such as this were subjected to a major academic revision in 2004.

grew to a majestic size, with some reaching 12 to 22 centimeters in length. The faunal evidence provided by the 52 trilobite species that have so far been uncovered from the nearly half-billion-year-old layers that run throughout the island is nothing less than astonishing. As anyone who has encountered them can readily attest, these striking Silurian remnants are uniformly large, beautifully preserved, and impressively three-dimensional—the stuff of any collector's dreams.

Prodigious specimens of *Diacalymene schucherti*, *Failleana magnifica*, and *Arctinurus anticostiensis* are often two or even three times the size of comparative species found in similarly aged Silurian horizons in nearby Ontario and New York. A dome-nosed *Failleana* recoverd in the 1960s measured an imposing 14 centimeters in length, and one of the island's few complete *Arctinurus* specimens checked in at an equally impressive 20 centimeters. When one adds the thick, coffee-colored calcite carapaces these trilobites exhibit to their remarkably lifelike preservation, you end up with Paleozoic prizes of the highest order.

"Early Silurian trilobites of Anticosti Island constitute something of a paleontological 'sleeper,'" said David Rudkin, formerly of Toronto's Royal Ontario Museum. "They're not nearly as widely known or appreciated as other Silurian trilobite faunas, like those of Dudley's famous Much Wenlock Limestone, New York's Rochester Shale, or even the Henryhouse Formation in Oklahoma. And that's a pity because in terms of diversity, preservation, and occasionally size, these are damned amazing trilobites."

Prospecting for fossils in this part of the world is not an especially new endeavor. As early as the mid-nineteenth century, scientific expeditions sponsored by the Geological Survey of Canada were exploring Anticosti's rugged cliff exposures, and the area's first described trilobite species emerged in 1859. At that time, and even today, seaward-facing sedimentary outcrops were found where the weathered-out layers of fossiliferous material were quite literally protruding from 20-meter-high cliff walls . . . or littering the beach below. Rarely has surface collecting been easier or more rewarding.

HADROMEROS NUPERUS (BILLINGS, 1866)

Early Silurian, Llandoverian (Late Aeronian—Early Telychian);
Jupiter Formation, Cybèle Member; Anticosti Island, Quebec,
Canada; 3.1 cm

This unusual trilobite was originally identified by scientists through disarticulated fragments. It is another genus that has direct evolutionary links to remarkably similar examples found in European rocks of the same age.

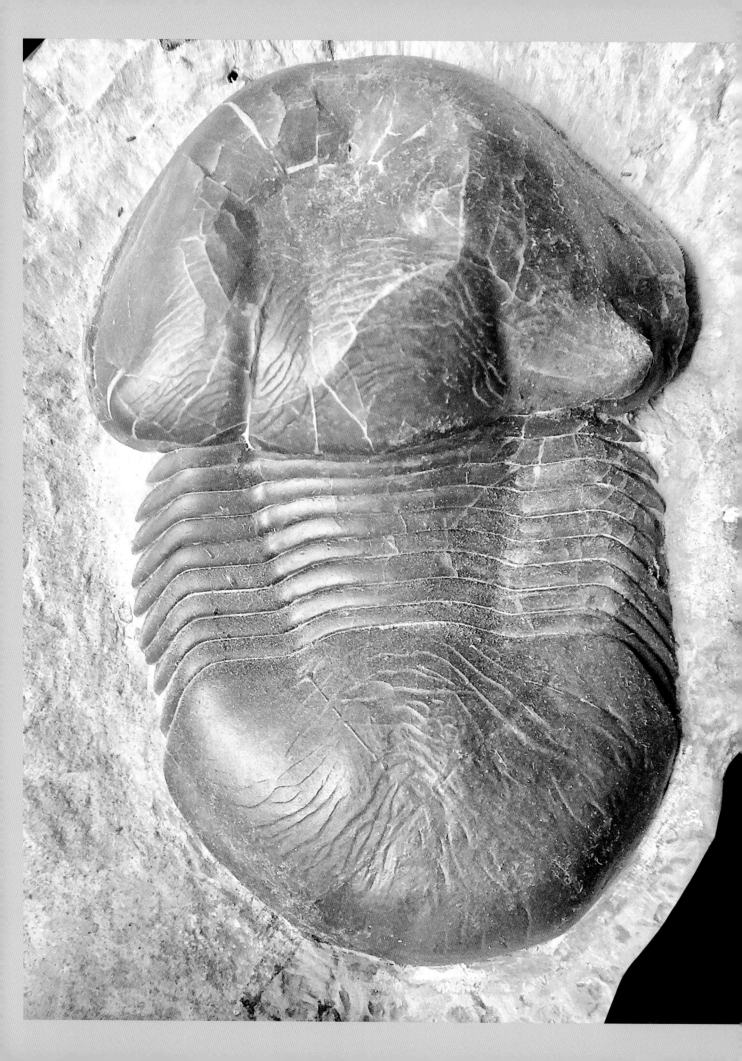

A scientific paper appeared in the mid-1860s in which nine initial Anticosti trilobite species were described (including examples of *Hadromeros* and *Scotoharpes*), and for a fleeting moment, it seemed as if this remote island was about to emerge as a burgeoning paleontological hotbed. Alas, it was not to be. Despite this promising early report chronicling Anticosti's trilobite riches, for whatever reason—most likely the island's distinctly out of the way setting—these fossils failed to capture the public's imagination in the same manner that other nineteenth-century trilobite discoveries had done, particularly those made in England and what is now the Czech Republic.

Another more complete monograph on Anticosti's Silurian fauna was presented in 1928, but even in the face of such renewed efforts, little heed was paid to the island's trilobite bounty. When that paper was unsuccessful in stirring up additional scientific interest or commercial support, this distant swath of land seemingly disappeared off the fossil destination map. Sporadic scholarly works appeared featuring Anticosti's ample trilobite caches throughout the 1970s and 1980s, and amateur collectors started visiting the island more frequently during that period. But as fossil strongholds such as Morocco and Russia took command of the world stage, the impressive trilobite fauna found on Anticosti's rocky shores was all but forgotten.

It wasn't until paleontologists Brian Chatterton (of the University of Alberta) and Rolf Ludvigsen (of the Denman Institute for Research on Trilobites)

visited this seaway-encircled location early in the twenty-first century, and subsequently presented an updated revision of those previous academic efforts, that Anticosti Island again fully entered the fossil lexicon. Once it did, however, that name—along with the magnificently preserved Silurian fauna it represented—quickly made impressive inroads into the paleontological psyche of trilobite enthusiasts around the globe.

"The variety of trilobites collected during our brief reconnaissance showed that the diversity was much greater than indicated by the existing literature," Chatterton and Ludvigsen wrote. "We returned to Anticosti on three separate occasions and accumulated extensive collections of trilobites from almost all the Silurian locations."

The results of these ambitious endeavors emerged in 2004 as the soft-cover volume, *Early Silurian Trilobites of Anticosti Island, Quebec, Canada*. In this book's densely illustrated 264 pages, Chatterton and Ludvigsen managed to add an abundance of new material to the island's mid-Paleozoic trilobite trove: 32 new species (to the existing 20) and one new genus (to the previous 29). Revealing black and white photographs featured such species as *Stenopareia grandis*, *Acernaspis orestes*, and *Nucleurus anticostiensis*, along with in-depth information regarding the island's singularly distinctive stratigraphy and paleoecology. Many of the trilobites were represented only by partial specimens, but the publication provided much needed insight on both the breadth and scope of the island's impressive fossil reserves.

Chatterton and Ludvigsen's work cast exciting new light on both Anticosti's trilobite fauna and the Silurian itself. Their writings also somewhat indirectly led to an influx of twenty-first-century amateur collectors eager to seek out their own share of the island's Paleozoic riches. This unexpected infusion of trilobite seekers initially alarmed Anticosti officials, many of whom did not fully comprehend

STENOPAREIA GRANDIS (BILLINGS, 1859)

Early Silurian, Llandoverian (late Aeronian—early Telychian); Jupiter Formation, Cybèle Member; Anticosti Island, Quebec, Canada; 8.2 cm

A very similar species characterized by having nine thoracic segments exists within the Late Ordovician rocks of Scotland.

CALYMENE SP. (AND CRINOIDS)

Early Silurian; Ellis Bay Formation; Anticosti Island, Quebec, Canada; largest trilobite: 3.1 cm

Here is an intriguing, captured in time Silurian moment. Dramatic fossils such as these provide a look at the interactions between species living along the Paleozoic seafloor.

how to handle adventurers who arrived without fishing rods or hunting rifles but with picks, hammers, and shovels. Understandably, those who control the non–national park elements of the island (Sepaq Anticosti is a provincial agency that oversees 60 percent of the island's landmass) were concerned that these paleontologically inclined efforts would somehow do irreparable damage to the area's natural beauty.

"You do sometimes need to keep an eye out for park officials," said a frequent Anticosti visitor. "It's not that collectors are doing anything wrong. It's more that the officials usually don't have a great grasp of what they're up to with their shovels and crowbars, so their first inclination is to believe they're somehow defacing the landscape. However, there is talk about the entire island becoming a protected UNESCO site, which could present more of an issue regarding any future trilobite expeditions."

Despite the flurry of digging activity that has taken place on Anticosti Island in recent years—perhaps spurred by the lingering "threat" of

UNESCO involvement and subsequent restrictions hanging over every collector's head—relatively few of the resulting trilobite specimens have managed to find their way onto the commercial market. This is particularly surprising when one considers that complete trilobites, especially of certain smaller *Acernaspis* and *Calymene* species (which usually average between 3 and 5 centimeters in length), are somewhat abundant within the island's outcrops.

In sharp contrast to the brisk flow to market activity enjoyed by both Ordovician and Silurian trilobite material pulled from noted exposures in Ontario and New York, prime specimens from Anticosti have proven to be almost as difficult to obtain as mythical trilobite teeth. In a notable fossil auction held in Montreal in the late 1990s, a limited number of large, well-preserved Anticosti trilobites (including a triple plate of a 10-centimeter *Diacalymene schucherti*) were quickly gobbled up by both museums and collectors. And in recent years, a few impressive examples of comparatively common species have sporadically appeared for sale on eBay. For the most part, however, first-rate specimens of the more unusual trilobites drawn from the island's rich sedimentary outcrops rank among the true "rara avis" items of the fossil world.

This scarcity may stem from the sheer difficulty involved in mounting a major exploratory expedition to Anticosti. It can cost thousands of dollars per person to travel to this lovely but inconvenient corner of the globe, transport in all the needed tools and provisions, and then stay for a week, or longer, in what often turns out to be an inhospitable climate. Another reason for the dearth of these attractive trilobites on the global market deals directly with the fact that those lucky enough to find unusual, top-quality specimens on the island tend to hold onto them like precious family heirlooms. And, quite honestly, a final reason you don't see many available Anticosti trilobites can be attributed to the overseeing eyes of the Parks Service, which continues to cast a generally negative eye toward any commercial outfit that might set foot on Anticosti's imposing coastline.

If you happen to find yourself face-to-cephalon with a trilobite culled from the rugged outcrops of Anticosti Island—whether in the field, a museum, or a private collection—perhaps you should take an extra moment to admire it. Odds are that you won't be seeing another example any time soon. This does little to detract from either the natural beauty or the scientific importance of these truly special trilobites, however, each of which provides invaluable insight into what life may have been like in those ancient Silurian seas.

RAPID REPORTS

Waldron Shale, Indiana

Rich sedimentary outcrops of Silurian age rank among the relative rarities of the fossil record. For trilobite enthusiasts, such well-known (and previously featured) formations as England's Wren's Nest and New York's Rochester Shale stand among the premier Silurian sites on the planet. Another notable, yet perhaps lesser-known U.S. locality lies within the Waldron Shale of Indiana, where beautifully preserved 425-million-year-old examples of such unusual trilobite species as *Metopolichas breviceps*, *Glyptambon verrucosus*, and *Trimerus delphinocephalus* have been found in outcrops that dot the southern half of the state. Perhaps closest in faunal content to the Rochester Shale, there are still marked differences in the material uncovered in the slightly younger strata of the Waldron, and the fossils here also appear to be less prevalent than in the New York locale. Because of their rarity, the attractive appearance of their calcite carapaces, and their lifelike preservation, Waldron trilobites are ranked as particularly prized specimens by hobbyists around the globe.

Henryhouse Formation, Oklahoma

Often overshadowed in the minds of collectors and academics by the impressive assortment of Devonian trilobites that spring from the shales of the neighboring Haragan and Bois d'Arc formations, the Silurian-age Henryhouse Formation is also a prolific source of prime Paleozoic material. Located amid the rolling Arbuckle Mountains in rustic Pontotoc County, Oklahoma, this site's 430-million-year-old trilobite fauna has long been treasured for its rarity, beauty, and scientific significance. Much like their world-renowned Devonian cousins, these Silurian Sooner State bugs are often three-dimensionally preserved in a thick, caramel-hued calcite, which contrasts strikingly against the surrounding buff-colored matrix. Indigenous species include *Dalmanites rutellum* (almost always found in an arched or enrolled state), the pustulose *Fragiscutum glebale*, the rare *Aanasobella asper*, and the formation's most common trilobite, *Calymene clavicula*, which can be 6 centimeters in length. Although its abundance and diversity has never proven equal to that of the legendary Haragan, the Henryhouse material has produced an impressive 11 trilobite species, encompassing 10 distinct genera.

Gotland, Sweden

On Gotland, a large island nearly 90 kilometers off Sweden's southeastern coast, trilobites have been excavated from the 420-million-year-old sedimentary rocks since 1851. Almost a millennium earlier, this 2,100-square-kilometer refuge served as an important Viking trading settlement, remnants of which can still be found hidden along the rugged shoreline. In key spots throughout Gotland, thick 420-million-year-old limestone beds have perfectly preserved a rich Silurian fauna featuring an abundant array of trilobite species. Usually presented in a fine, toffee-colored calcite, these wonderfully inflated, although generally diminutive, specimens include such trilotypes as *Proetus granulatus, Kettneraspis angelini, Sphaerexochus latifrons*, and *Calymene neotuberculata*. Often compared in both age and preservation to the renowned Much Wenlock outcrops of central England, Gotland specimens are perhaps even more prized due to their relative scarcity. With many of the fossil-bearing Gotland layers now completely submerged under the waters of the surrounding Baltic Sea, recently found examples from this locale have been few and far between.

SILURIAN TRILOBITE PHOTO GALLERY

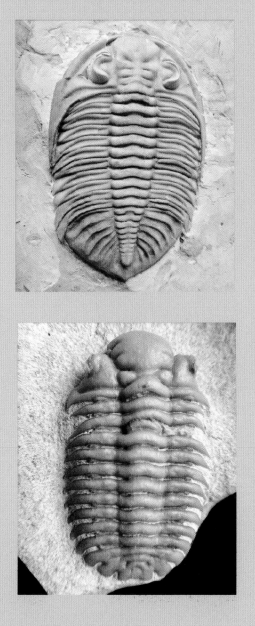

(OPPOSITE PAGE) *DIDREPANON SP.*

Late Silurian; Bannisdale Slate Formation; White Howe, Southern Lake District,; Cumbria, England, UK; 18.2 cm

This is one of the largest Silurian cheirurid trilobites ever found in Europe. Legend is that it was discovered within a large stone block that had been used to construct a wall in northern England.

(ABOVE, TOP LEFT) *DALMANITES RUTELLUM* (CAMPBELL, 1967)

Silurian; Henryhouse Formation; Oklahoma, United States; 6.2 cm

This is one of the few prone, dorsal examples of this rare bug from the American Southwest. Other known specimens tend to be either arched or totally enrolled.

(ABOVE, BOTTOM) *ANASOBELLA ASPER* (CAMPBELL, 1967)

Middle Silurian; Henryhouse Formation, Hunton Group; Pontotoc County, Oklahoma, United States; 5.1 cm

This midsized cheirurid rates as one of the rarest finds from the Henryhouse.

(ABOVE, RIGHT) *PROROMMA SP.* (THOMAS, 1981)

Middle Silurian; Lower Wenlock Series, Shinewoodian; Buttington Mudstone Group, Centrifugus Biozone; Former brickworks quarry, Buttington, Shropshire, England, UK; 5.3 cm

This attractive specimen was found in a quarry that has proven to be the primary source for these colorful European cheirurids.

(ABOVE) KETTNERASPIS CRENATA ANGELINI (PRANTL AND PRYIBYL, 1949)

Middle Silurian, Wenlockian; Klinteberg Formation; Djupvik, Gotland, Sweden; 2.2 cm

This small but attractive specimen displays the thick calcite preservation that has made Gotland trilobites coveted by collectors around the globe.

(OPPOSITE PAGE) METOPOLICHAS BREVICEPS (HALL, 1864)

Silurian; Waldron Shale; Shelby County, Indiana, United States; 6.1 cm

This is generally considered to be the only complete example yet found of this impressive midwestern lichid.

(ABOVE) *OMMOKRIS VIGILANS* (HALL, 1861)

Lower Silurian, Niagaran Series, Medina Stage; Edgewood Formation; Grafton, Illinois, United States; 4.2 cm

This telescope-eyed species represents one of the most amazing—and rarest—trilobites emerging from the famed dolomitic deposits of Grafton.

(LEFT) *TRIMERUS DELPHINOCEPHALUS* (GREEN, 1832)

Silurian; Waldron Shale Formation; Waldron, Indiana, United States; 14 cm

This genus is also found in New York, Bolivia, and England, but a complete specimen from this locale is highly unusual.

(OPPOSITE PAGE) *ENCRINURUS EGANI* (MILLER, 1880)

Silurian; Niagaran Group; Lemont, Illinois, United States; 4.2 cm

Note the stalked eyes atop the cephalon of this dolomitic specimen. Such a morphological feature may have allowed the trilobite to burrow into seafloor ooze while maintaining visual vigil on the surrounding world.

(ABOVE) CALYMENE CLAVICULA CAMPBELL, 1967

Silurian; Henryhouse Formation; Pontotoc County, Oklahoma, United States; 6 cm

This is a large, impressive example of a once common trilobite. In recent years, little exploratory work has been done in the Henryhouse, making top-notch specimens more difficult to obtain.

(OPPOSITE PAGE) SPATHACALYMENE NASUTA (ULRICH, 1879)

Silurian; Osgood Formation; Ripley County, Indiana, United States; 9.8 cm

With its pronounced rostrum, this species ranks among the most unusual trilobites in the world. This is a particularly large example of a species that often appears atop the "want" list of serious trilobite enthusiasts.

(ABOVE) *SENTICUCULLUS ELEGANS* (XIA, IN CHANG, 1974)

Lower Silurian; Xiushan Formation; Western Hunan Province, China; 6.1 cm

One of the most visually striking Silurian species in the fossil record, this unusual trilobite has been found complete only on rare occasions.

(OPPOSITE PAGE) *DICRANOPELTIS SCABRA PROPINQUA* (BARRANDE, 1846)

Silurian; Motol Formation; Lodenice, Czech Republic; 4.4 cm

The cranadia of this attractive and rare specimen was torn away during extraction, exposing the hypostome. This genus is also found in New York and England.

THE DEVONIAN PERIOD

419–359 Million Years Ago

MOROCCAN TRILOBITES:
TREASURES FROM THE DESERT

T O REACH ALNIF, A SMALL, dusty town located 650 kilometers southeast of Casablanca, you first head to the middle of nowhere . . . then make a left. A hearty eight-hour drive from Morocco's capital city over a precarious two-lane road that bisects the imposing Atlas Mountains, Alnif seems a safe bet to avoid any future reputation as a major tourist destination. Average daytime temperature in this rural village (population 3,500) is a scorching 90° Fahrenheit (32° Celsius), and although most residents seem oblivious to the heat, the weather often serves to bring any local action to a grinding halt. Activity in Alnif is conducted at two speeds, slow and slower, and those unwilling or unable to adapt to the town's tempo will quickly find themselves hopelessly out of step.

Why would anyone want to visit this remote hamlet on the edge of the Sahara Desert where until recently the main means of transportation was burro-driven cart and the only apparent social enterprise was drinking mint tea? Sure, the occasional Hollywood movie (including everything from the classic *Lawrence of Arabia* to the instantly forgettable 2005 action flick, *Sahara*) has been filmed in the nearby desert, with Alnif serving as pre- and postfilming headquarters. But there's got

BASSEIARGES MELLISHAE CORBACHO AND LÓPEZ-SORIANO, 2013

Middle Devonian, Eifelian; Bou Tchrafine Formation; Jorf, Tafilalt Region, Morocco; largest trilobite: 3 cm

Here is an amazing colony of ornate lichids from a recently opened dig site located near the town of Jorf. Trilobites were apparently highly social creatures, often living in closely packed groups such as this.

to be more to it than that. Once one delves slightly under the fine layer of sand that pervades everything in and around this town still locked very much in the heart of nineteenth-century ideals, it becomes quickly apparent that there is indeed something more.

If you're a trilobite collector, the lure of Alnif is obvious. Over the past four decades, the city has emerged as a key hub in Morocco's growing fossil trade. During that time, a dizzying variety of vertebrate and invertebrate material—and trilobites, in particular—has helped establish both the town and the nation among paleontology's most renowned hotbeds. A stroll down Trilobite Alley—a dark and somewhat mysterious thoroughfare of well-packed dirt in the heart of Alnif's "business district"—is like taking a step back in time. The small, generally well-maintained, two-story houses that line the narrow passage would have looked much the same a hundred or two hundred years ago. Women, when they are seen at all, remain in the shadows, covered from head to toe in flowing black robes, strictly following Muslim law. Bands of laughing children roam the streets asking for candy with one hand while swatting away swarms of ever-present flies with the other.

There's no question that Alnif is a man's town. It's a rough and tumble place where life is far from easy, and residents have developed a razor-sharp survival sense that enables them to quickly and effectively deal with just about any situation that may arise. And it is unquestionably the local trilobite dealers who proudly sit atop the city's paleontological pecking order. These men control the dozen shops that line Trilobite Alley, some of which are little more than one-room shanties with bare light bulbs illuminating hundreds of specimens that either fill large wooden display cases or are scattered haphazardly across the area-rugged floor.

The absence of an upscale showroom ambience is quickly forgotten upon confronting the diversity and rarity of the trilobites strewn about. For the true bug enthusiast, a feeling of trilobite overload can easily take hold with the sheer volume of Paleozoic product threatening to inundate the senses. No matter where one chooses to look, dozens of Devonian species—some partially exposed, others still almost fully encased within a slate-gray sedimentary matrix—sit edge to edge with less common Ordovician and Cambrian examples.

None of these perpetually smiling Alnif merchants seem even slightly interested in separating themselves from the local competition by specializing in premium trilobites drawn from one geological period, or from a solitary sedimentary deposit. Rather, each shop appears to feature almost an identical stock of Moroccan rock, the only obvious differences being in the quantity and quality of their fossiliferous holdings—along with the manner in which those specimens are presented. On occasion, particularly unusual trilobites, or those that have been more carefully prepared by air-scribe methods, are proudly displayed behind somewhat flimsy plastic showcases, whereas the more pedestrian material is piled like stone kindling in a dim corner.

Genera familiar to all trilobite collectors—*Phacops*, *Asteropyge*, *Harpes*, and *Leonaspis*—are all there. They vie for attention alongside more exotic species such as *Psychopyge elegans* (featuring a 5-centimeter-long nose "sword"), *Radiaspis sp.* (with its characteristic rows of long, pointed spines), and *Dicranurus monstrosus*, a particularly elegant trilo-type highlighted by a dramatic combination of ramlike "horns" atop its head and

THYSANOPELTELLA N.SP.
Middle Devonian; Eifelian Stage; Morocco; largest trilobite: 7.1 cm

This is an intriguing group of as yet undescribed scutellid trilobites. They were found in a still "secret" desert-adjacent location where the rock features both an unusual texture and color.

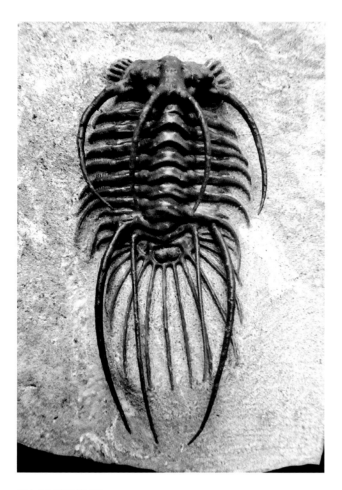

RADIASPIS SP.
Middle Devonian, Eifelian; El Otfal Formation; Jebel Ofaténe, Morocco; 6.6 cm

Science is still trying to determine the role the amazing proliferation of spines on this species may have played.

long, flowing spines along its body. There are even attractive multiplates, some featuring as many as a half-dozen complete trilobites that often represent two or more distinct species.

Every year a horde of new trilobite material—uncovered from increasingly remote Moroccan quarries—leaves even the most seasoned Alnif dealers confused and confounded. Some of these specimens hail from locations that skirt the steaming desert. Others are culled from Devonian dig sites that sit mere kilometers from downtown Marrakesh. A visitor may ask any of the experienced

town merchants about the genus and species of these strange new creatures in either of the city's spoken languages, French or Arabic, but often the best translation of their shrugging response is "your guess is as good as mine."

Their lack of academic insight does little to deter these well-schooled shop owners from instantly gauging an unfamiliar specimen's possible value . . . especially as soon as visiting eyes widen in obvious appreciation of its rarity. Few of the displayed trilobites carry noticeable tags indicating identity or cost, so Moroccan merchants are free to engage in their favorite business activity—haggling over prices with potential customers. That practice is sure to commence as soon as an offer is made on any of their half-billion-year-old wares. No matter what an Alnif trilobite dealer may initially answer when asked about the price of a particular specimen, he *expects* the prospective purchaser to come back with a counteroffer—and he'll be sorely disappointed, and somewhat annoyed, if that customer fails to do so. Indeed, it seems to matter little how high or low an initial offer may be; without the expected negotiating game, even the most dedicated Alnif dealers seem to quickly lose interest in the transaction.

"Moroccan merchants have been honing their business skills for centuries," said Bill Barker, one of the preeminent American importers of Moroccan fossils, and someone who lived in the country for over a decade. "It's been part of their culture since the old 'caravan' days. They've quickly learned about trilobites, and they've become very adept at carrying on conversations about them with scientists and collectors from around the world. But, to be honest, to them it really doesn't matter what they're selling; it could be rugs, meteorites, or fossils. For most of them, it's all just business."

Once or twice each week the Trilobite Alley merchants leave their Alnif shops. They hop into their well-worn, European-model cars (among the

few motorized vehicles in town) and make the two-hour drive over pothole-laced back roads to reach the local fossil quarries. There, teams of hired field-workers—many of whom are members of the dealer's extended family—dig alongside roaming bands of nomads, a distinctive bunch who maintain strict independence from any individual fossil dealer. From sunup to sundown, day in and day out—although some do take off Friday, the traditional Muslim day of rest—this dedicated contingent of dust-covered laborers patrols the surrounding hillsides in search of potentially valuable Paleozoic treasures. They relentlessly attack the surrounding rock surfaces with everything from sharpened pieces of scrap metal to fancy Swiss-made rock hammers, and each tool proves equally effective in the hands of this resourceful crew. Once they find the trilobite bounty they've been seeking, they do their best to extract it—any which way they can.

The Middle Cambrian *Acadoparadoxides levisetti* formations are excavated in a relatively sophisticated, if scientifically problematic, manner. Teams of well-trained quarrymen use controlled dynamite charges to loosen layers of sedimentary strata. When the telltale evidence of

DROTOPS ARMATUS HOPLITES STRUVE, 1995

Middle Devonian (basal Givetian); Bou Dĩb Formation, Lower Member; Jebel Issoumour; Maïder Region, near Alnif, Morocco; 15.2 cm

Perhaps the most recognized species emerging from the Moroccan trilobite "invasion" that has taken place since the 1990s, this large trilobite appears armed like a primordial battle tank.

Photo courtesy of Martin Shugar

trilobites is spotted, the fossil-laden rock is then forcibly removed from its enveloping matrix and carted down the mountainside by burro. There, it is handed over to workers waiting in preparation "factories," most of which are little more than small shacks lacking both electricity and plumbing. It is in these basic but efficient facilities that the large trilobites are made ready for sale. This is accomplished either by chipping them out of their surrounding stone with a primitive nail and hammer method (if they are relatively complete specimens) or by somewhat haphazardly puttying together often mismatched morphological pieces in a ham-fisted attempt to make them more presentable for future transactions. Some examples are sent to another wing of the factory where uniquely skilled artisans carefully carve missing parts of the various trilobites (usually free cheeks and genal spines, but occasionally entire heads or tails) into the surrounding rock.

In contrast, the more delicately constructed Devonian *Dicranurus monstrosus* trilobites are mined by teams of wandering nomads from kilometers-long trenches. Over the last few decades, these 2-meter-deep channels have been gouged into the hard limestone overburden by hand-wielded tools, creating a crazy, zigzag pattern that follows the topographic profile of the surrounding mountains. Once a promising rock has been detached from these ragged, deep-ditch excavations, the colorfully clad natives painstakingly break apart and then lick (yes, lick, water can be a scarce commodity in these parts) every exposed

surface, a process that allows them to better view their elusive trilobite targets. The added moisture makes the dark, charcoal-colored calcite of the trilobite fossil contrast more prominently against the uniform, mid-gray color of the surrounding matrix. Sometimes, if they're lucky, a detailed cross-section of hard to distinguish trilobite segments appear, which is often enough to alert the diggers' well-trained eyes that a tantalizing piece of their highly sought prize has been revealed. These finds usually produce disarticulated or incomplete specimens, but occasionally a true mid-Paleozoic gem is uncovered—a complete example of a rare trilobite species.

Invariably the Alnif dealers know just where these field-workers are located, although reaching them amid the growing piles of discarded sedimentary rubble can prove challenging. With patience, a bit of cajoling, and the necessary transfer of funds, in a single afternoon the merchants can acquire enough potentially complete specimens to keep the men who prepare trilobites for their shops busy for the next few days.

Over the last decade, a steady influx of state-of-the-art pneumatic scribes and air-abrasive drills has arrived in these remote North African regions, along with a corresponding array of sturdy gas-powered generators to power them. When placed in the hands of talented Moroccan technicians, these tools expedite the preparation process, and to the delight of collectors around the globe, this has markedly improved the overall quality of the resulting trilobite specimens.

"It's a very effective system," Barker said. "The diggers dig, the preppers prep, and the merchants control it all. They're each independent yet interdependent, and all this activity centers on those times when the Alnif trilobite dealers go to acquire new material."

Such weekly desert outings have been occurring with regularity over the last 40-odd years, ever

LICHAS MAROCANUS (DESTOMBES, 1968)
Ordovician; Ktaoua Formation; Tazarine, Morocco; largest trilobite: 11.3 cm

This is an attractive—and rare—assemblage of large lichid trilobites. Note the single ventral specimen, which indicates that these may all represent molted exoskeletons.

EUDOLATITES SP.
Lower Ordovician; Ktoua Formation; El Kaid Errami, Morocco;
21.5 cm

A layer of these large trilobites was discovered in 2014 and was quickly exhausted. Perhaps a dozen complete specimens were recovered.

have numbered in the dozens—annually produce hundreds of the relatively common phacopid species (also known by genus names such as *Drotops, Geesops, Austerops*, and *Smoothops*) and yield perhaps as many as 20 examples—and as few as one or two—of the rarer species.

Interested observers have speculated that only a small fraction of Morocco's available trilobite resources have been fully explored. Although most of the quarries first mined in the 1980s and 1990s have now been tapped out, fresh Paleozoic dig sites are still being discovered every year. These new excavations include an exciting Middle Devonian quarry located near the centrally situated town of Jorf. This site first rose to prominence in 2016 with the emergence of perfectly preserved examples of the rare lichid genus *Acanthopyge*. Since then, the number and variety of Jorf trilobites—many featuring a distinctive, vitreous preservation—have continued to expand, with most quickly finding their way into the appreciative hands of both the collecting and academic communities.

"The Moroccans take a great deal of pride in not only their collecting abilities but in their growing scientific understanding of what they're doing," Barker said. "That's why you're always going to see new trilobites come from different locations throughout the country. It's partially about the money . . . but it's also about the satisfaction of finding something they haven't seen before."

In recent years, more and more of these unusual North African trilobite species have begun infiltrating the commercial fossil scene. Increasingly exotic examples, many featuring rows of freestanding spines and intricately patterned surface textures, continue to show up in surprisingly large numbers at leading trade shows in Tucson, Denver, Tokyo, and Munich. Dozens of dealers from Germany, Japan, the United States, and England, in addition to the expected Moroccan crews, set up shop at these events, offering scores of spectacular

since Berber tribesmen living in the outskirts of Alnif first discovered the "Magic Mountain" that yielded the region's initial supply of Devonian-age trilobites. No one knows exactly how many specimens have been uncovered in Morocco's vast and varied fossil fields since then, and no one will even warrant a guess regarding how many trilobites have subsequently been bought and sold by the Alnif merchants. When pushed on the subject, one dealer estimated that the numerous Devonian trilobite quarries—which at various points in time

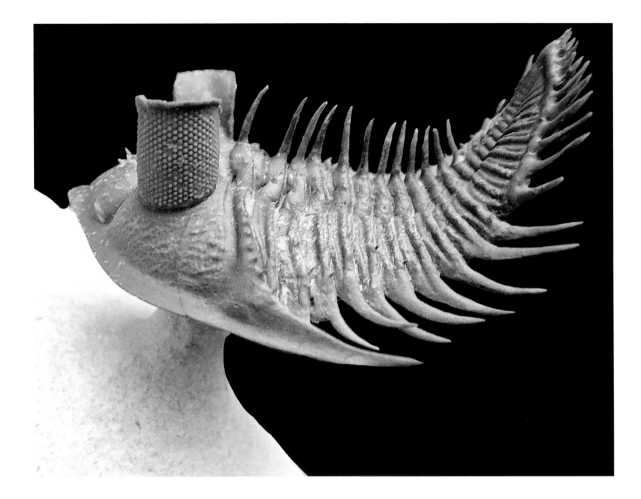

specimens—often at equally spectacular prices. Yet as Morocco has assumed a leading position among the planet's most renowned and revered paleontological storehouses, troubling questions have continued to fester in certain circles. Many speculate about the true nature—and authenticity—of at least some of these supposedly unsullied Paleozoic remnants.

It is undeniable that there are currently more Moroccan trilobites available on the global market than ever before, including an increasing number of exceptionally rare species. But is this invertebrate influx based on a sudden surge of Moroccan field-workers seeking out previously unknown or underworked quarries? Or, as some concerned parties have speculated, have groups of talented Alnif artisans turned to blatantly fabricating a variety of marketable specimens—including anatomically accurate examples of *Selenopeltis, Odontochile,* and *Ceratarges*—with the intent of easily increasing their income?

Even top collectors, along with a growing number of museum curators and university professors, have admitted to being fooled at one

ERBENOCHILE ERBENI
(ALBERTI, 1981)

Devonian; Timrhanrhart Formation; Jebel Issomour, Morocco; 6.2 cm

This ranks among the world's most unusual species. It features some of the largest compound eyes to be found in the trilobite world and has a shielding "brim" atop each eye column.

time or another by first-rate Moroccan handiwork. These duplicitous efforts have effectively utilized highly detailed trilobite casts—comprised of everything from cheap automotive plastic, to top-quality resin, to a rough mud-based cement—placed on real matrix, to create a line of faux fossils that have infiltrated the international marketplace right alongside the genuine Paleozoic articles. To put it simply, a growing and often contentious debate has erupted over whether many Moroccan trilobites are true relics of the distant past or mere artful concoctions.

"There's no doubt that some of the trilobites coming out of Morocco are at least partially fabricated," said Barker. "There is a school of workers who have taken to molding certain Devonian trilobites out of plaster or plastic, and then placing them on actual rock. But they insist that is done more for tourists than anyone else."

Back on the sun and sand swept alleys of Alnif, such debates seem to be having minimal impact on the community's approach to marketing their fossil goods. The dealers are more aware than ever of the differences between fake trilobites and real ones, but in their heart of hearts, it's not a question of science but of commerce. After all, money is money. In a town where an education is often limited to what one learns on the streets, arguments presented by North American and European college professors, or even well-intentioned amateur collectors, tend to fall on deaf ears. However, these merchants have begun to realize that as more and more people become aware of both the wonders and the inherent problems associated with the fossil trilobites surfacing from the majestic mountains of Morocco, additional focus will be placed on weeding out authentic specimens from those possessing a more questionable geological pedigree.

Should the typical fossil collector be wary of acquiring Moroccan trilobites from local dealers at area rock and mineral shows or on an ever-present outlet like eBay? Probably not. A relatively sharp eye and a basic knowledge of trilobite morphology should eliminate most potential difficulties, although in recent years the sophistication in trilo-fakery has often grown step for step with any increase in collector savvy.

The fact remains that nowhere else on Earth are more spectacular and fascinating *legitimate* trilobite specimens coming to light than in the rich sedimentary soils of Morocco. For a collector or a scientist to turn his or her back on such a proven source of prime Paleozoic material would appear to be both foolish and unnecessary. As one Alnif fossil dealer proclaimed as he hawked his wares outside of his tiny shop, "These trilobites are the best in the world." But in the words of consumers throughout the ages, "Let the buyer beware."

TRILOBITE ENROLLMENT

There is an apocryphal tale that has become part of trilobite folklore. It tells of how back in the early 1970s, a Berber tribesman living on the edge of Morocco's Sahara Desert was walking near the mountains that encircled his home. As he was strolling past one imposing 2,000-meter-high peak, he heard the distressing sound of falling rocks and noticed a fist-sized piece of debris as it began cascading in his direction down the perilous slope. When that chunk of sedimentary stone had come to rest, he bent to investigate what the mountain had yielded. Much to his surprise, he found a perfectly enrolled, ball-shaped Devonian-age *Drotops* trilobite lying at his feet. The unexpected discovery intrigued the Berber, and he sought the spot from which this strange rock had fallen. For weeks on end he searched, climbing and exploring, finally locating the fossil-bearing strata on the so-called Magic Mountain that would soon ignite the entire Moroccan trilobite explosion.

This tale serves as a somewhat roundabout means of broaching the subject of trilobite enrollment and presenting some very basic questions: Why did these primitive arthropods enroll, and when did they first begin doing so? Of course, many modern animals, from isopods to armadillos, enroll for safety, and we can assume that trilobites used this defensive stance for much the same reason. With their thick, often spine-encrusted calcite shells, an enrolled trilobite could present quite the daunting challenge for any predator seeking an easy meal. And if that alone wasn't a good enough reason for these creatures to enroll, scientists have recently speculated that the compact, spherical form provided by enrolling also may have aided some trilobite species in battling unfavorable conditions during the heights of primordial oceanic storms.

During enrollment, the flexibility of its thoracic segments allowed a trilobite's rigid cephalon and pygidium to interlock, encasing the creature's delicate internal organs and soft ventral appendages. Academics

***PARAHOMALONOTUS CALVUS* CHATTERTON, FORTEY, BRETT, GIBB, AND MCKELLAR, 2006**

Devonian; Timrhanrhart Formation; Jibel Gara el Zguilma; Morocco; diameter: 10.3 cm

Those who have seen this specimen often refer to it as the "baseball bug" due to its size and near perfect enrollment.

PEDINOPARIOPS (HYPSIPARIOPS) VAGABUNDUS STRUVE, 1990

Middle Devonian, Eifelian-Givetian Boundary; Bou Dib Formation; Maïder Region, Morocco; diameter: 6.4 cm

Much like modern armadillos, it is generally believed that trilobites enrolled for protection from predators.

will assuredly state that there were various types of trilobite enrollment—including Sphaeroidal (when all the thoracic segments flexed as part of the enrolling procedure) and Discoidal (when only the anterior segments of the thorax flexed)—but the bottom line is that each served the same basic purpose of protecting the host trilobite from hostile situations. The fossil record indicates that a preponderance of the 25,000 recognized trilobite species could enroll and that it was a common occurrence among virtually all orders of these unique invertebrates.

It is known that this feature lasted until the demise of the trilobite line at the tail end of the Permian, but when did this notable evolutionary development begin? A few Cambrian trilobites

appear to have possessed the ability to enroll, although the rare fossilized remains of such specimens seem more "folded" than truly enrolled. However, by the Lower Ordovician, some 470 million years ago, trilobite enrollment was already quite advanced. Many of the magnificent specimens emerging from Russia's famed Ordovician-age Asery Horizon quarries are found in various states of enrollment, often presenting the appearance of nearly symmetrical fossil balls. And by the Devonian, perhaps the most renowned of these trilobite "rollers" had emerged. These are the beautifully preserved *Eldredgeops* species from Ohio, where their 400-million-year-old circular conformity is broken only by the pronounced protrusions made by their compound eyes.

Judging by all the fossil evidence, it is apparent that the ability to enroll played a vital role in helping trilobites not only survive but flourish, allowing them to dominate the planet's seas for more than a quarter-billion years.

TRILOBITE PATHOLOGY AND PREDATION

Judging by the gruesome assortment of lethal-looking bite marks and healed injury scars that frequently adorn their fossilized exoskeletons, the daily existence of a trilobite seems to have been filled with formidable trials and tribulations. The graphic evidence of their hazardous lifestyle offers rather emphatic proof that the primal seas in which these Paleozoic arthropods flourished for nearly 300 million years provided neither a hospitable environment nor a safe path through evolution's oft-treacherous labyrinth. Unquestionably, those ancient oceans were infused with an ever-changing and an ever more dangerous cast of predatory characters, all seemingly intent on turning the local trilobite population into tasty midday morsels.

***GLOSSOPLEURA GIGANTEA* (RESSER, 1939)**

Middle Cambrian; Langston Formation, Spence Shale Member; Antimony Canyon, Utah, United States; 13.7 cm

Note the large bite mark on the right side of this trilobite's body.

ELLIPTOCEPHALA SP.

Lower Cambrian; Rosella Formation, Atan Group; British Columbia, Canada; 13.4 cm

The jagged bite mark adorning this specimen's thorax provides ample evidence of the dangers present in the Paleozoic seas.

From the moment trilobites emerged on the global scene some 521 million years ago, they were placed in continual jeopardy by a lethal legion of increasingly ravenous creatures ranging from the legendary Cambrian arthropod *Anomalocaris canadensis*, to jet-propelled Ordovician cephalopods, to armor-plated Devonian fish. Current theory suggests that some trilobites, such as the large Cambrian genus *Olenoides* (which could exceed 15 centimeters in length), may even have taken an occasional cannibalistic turn upon some of their smaller trilobite brethren. No matter which blade-mouthed behemoth may have unleashed the most savage attacks upon them, fossilized trilobite carapaces sporting the remnants of either healed or potentially fatal bite marks are pervasive within the fossil record. Indeed, in certain locales, such as the famed Middle Cambrian fossil beds of Delta, Utah, jagged, monster-made incisions can often be seen adorning the calcite shells of the region's abundant *Elrathia kingii* and *Asaphiscus wheeleri* specimens.

By the dawning of the Ordovician, 485 million years ago, a preponderance of trilobite species had developed the ability to enroll, which provided a significantly greater degree of protection from the myriad hostile inhabitants of the marine world in which they lived. Other trilobites had evolved to grow rows of menacing spines covering their carapaces, which afforded additional defense from predators. Still others generated increasingly thick calcite shells with each successive molt, providing these species with a greatly enhanced chance to survive and pass along their genetic advantages. Despite their best efforts to shield themselves, however, the grim pathological evidence revealed in sedimentary stones around the globe indicates one thing—trilobites were still often the victims of fierce predacious attacks.

Although scientists have yet to offer an irrefutable reason for this, most of the bitten trilobite exoskeletons appear on Cambrian-age and Ordovician-age fossil specimens. Yet even in the Silurian, Devonian, and Carboniferous, the latter a time during which the entire trilobite line was already in steep decline, examples from sites in Morocco, Russia, Canada, and the United States demonstrate detailed proof of their carapace-cracking injuries, many of which the trilobites apparently survived. In fact, even on some of the most blatantly bitten shells in the fossil record, evidence exists of regenerated asymmetric spines. This suggests that the trilobite not only managed to make it through another day but may have continued living in those primal seas for a considerable length of time.

Trilobites—much like modern arthropods—frequently shed their hard, protective shells, so their damaged exoskeletons could not begin to heal until their subsequent molt occurred. Occasionally, the mineralized edges of these next molt bite marks appear surprisingly smooth, a sign that the regenerating process had already begun. Often, however, the serrated yet symmetric pattern on the fossilized trilobite shell indicates something far more dire—that the initial injury may have indeed proven fatal.

Despite ongoing academic efforts, attributing these half-billion-year-old wounds to any explicit attacker remains a Deep Time mystery. In some cases, especially during the Cambrian, an abundance of those marks are of a distinctive shape and appear to have been caused by a specific marine menace. One likely candidate is the previously mentioned *Anomalocaris*, the reputed terror of the Cambrian seas, whose imposing appendages and razor-edged mouth plates have been found in close association with trilobites everywhere from British Columbia, to Australia, to Southern China. Some paleontologists have recently begun to question whether an *Anomalocaris* mouth plate was powerful enough to crack a thick trilobite carapace, but until a better carnivorous candidate swims by, these "abnormal shrimp" rank among the prime suspects in perpetrating these violent Paleozoic assaults.

Whether or not trilobites managed to survive these ferocious attacks, it is clear that their perceived pain served as our eventual gain. Unquestionably, the discovery of these flagrantly injured fossilized carapaces has provided both scientists and collectors with an additional portal through which to view life in the Paleozoic oceans. After all, it's one thing to examine 500-million-year-old fossils featuring the detailed impressions of ancient life-forms. It's quite another to be able to interpret those impressions in a manner that provides greater understanding of what the battle for daily existence may have been like during those long-gone yesterdays when trilobites filled the seas.

TRILOBITES OF BOLIVIA: *SKY HIGH IN THE ANDES*

Trilobites were perhaps the most global of Paleozoic inhabitants. During their 270-million-year span of existence, these fascinating invertebrates managed to occupy just about every ocean-covered corner of our planet. From warm tidal bays to sun-dappled undersea plateaus, from cool offshore shelves to yawning deep-water coves,

trilobites took full advantage of this bountiful profusion of ecological niches, continually showing themselves to be a particularly versatile and resourceful animal class.

Because of this adaptability—along with the subsequent "push" provided by the continent-shifting forces of plate tectonics—the fossilized exoskeletons of trilobites have been discovered everywhere that sedimentary outcrops of the proper geologic age occur. For academics and collectors, however, a select number of trilobite-rich sites stand above and apart from the rest. Whether due to the exceptional preservation of their fauna, the diversity of their inherent species, or the inhospitable nature of their current location atop the global lithosphere, these prime paleontological outposts often offer distinctive perspectives on the distant past, as well as testing anyone's willingness to travel to the world's farthest recesses in their quest for these marvelous marine remnants.

One of the most intriguing and out-of-the-way destinations ever to yield a scientifically significant bounty of complete trilobite material can be found atop the high and dry Altiplano—the legendary plateau located in the outskirts of La Paz, the capital city of Bolivia. There, up

BOULEIA DAGINCOURTI (ULRICH, 1892)

Middle Devonian; Belen Formation; La Paz, Bolivia; 3.2 cm

This attractive specimen displays both sides of the positive/negative concretion that traditionally house the trilobites that emerge from this high-altitude repository.

TRIMERUS (DIPLEURA) DEKAYI BOLIVIENSIS WOLFART, 1968

Middle Devonian; Belen Formation; La Paz, Bolivia; 6 cm

This genus can be found worldwide, including locations in England and the United States. It appears in both Silurian and Devonian exposures.

in the Andes Mountains, at a breathtaking altitude of more than 3,600 meters, closely aligned formations that include the Sicasica, Belen, and El Carmen have provided a series of prolific outcrops brimming with Devonian-age trilobite fossils. These unusual specimens often emerge from their 370-million-year-old burial grounds contained within small, mud-colored limestone concretions that are generally less than 7 centimeters in length. When these egg-shaped nodules are carefully cracked open, they occasionally reveal positive/negative fossil splits that display outstanding preservation and detail.

Fossil-bearing concretions are a relatively common geological phenomenon, appearing in Paleozoic layers everywhere from the eastern United States, to the English Midlands, to the desert-hugging trilobite beds of North Africa. They occur when mineral-rich elements housed within the sedimentary strata attach to and then condense around a static object, such as a trilobite carapace. But in Bolivia, these concretions are particularly prolific, both in their numbers and in their degree of fossiliferous content. Some estimates indicate that over half the concretions found in Bolivia's sky-high Devonian formations (which, for those wondering, first rose to their impressive elevations during the Cretaceous Period, some 120 million years ago) contain the remains of at least some sort of organic material, although, unfortunately, only a few feature complete trilobite specimens.

"Bolivian trilobites enjoy a special form of preservation," said Juan Colon, a La Paz native who has prospected for South American trilobites for more than two decades. "But even after you find them, you need to be very skilled to break the concretion properly and reveal the contents."

For many years, right up until the early 1950s, virtually all that the world knew of these extraordinary, three-dimensional South American trilobite concretions had been garnered from poorly preserved specimens sold at local markets by indigenous Bolivian merchants. Prior to that time, there seemed to be minimal academic interest—or collector curiosity—regarding these unusually preserved Paleozoic relics. That attitude slowly began

MALVINELLA BUDDEAE LIEBERMAN, EDGECOMBE, AND ELDREDGE, 1991

Middle Devonian; Sicasica Formation; La Paz, Bolivia; 5.1 cm

This is an elegant example of a species commonly found within Devonian-age Bolivian concretions. These nodules occur when mineral-rich mud forms around a static object, such as a trilobite carapace.

BELENOPYGE BALLIVIANI PEK AND VANEK, 1991
Middle Devonian; Belen Formation; La Paz, Bolivia; 4.5 cm

Somewhat similar examples of the *Belenopyge* genus have been found in Devonian outcrops throughout Oklahoma and Morocco.

At the same time, it was possible for travelers to stroll down a back alley in the vibrant city of La Paz, which boasts a population of almost one million, and see Devonian-age trilobites for sale, often for only a few centavos, at street corner stands or in local shops. Within these offbeat emporiums an impressively diverse display of concretion-encased fossil relics could frequently be found hiding among the colorful native handiwork and for-tourists-only oddities—which on occasion even included mummified llama and alpaca remains. Many of the fossils were partial trilobite specimens, lacking either their head or tail, but others were complete weather-worn examples of the most commonly found species. On occasion, however, through luck, happenstance, or sheer persistence, true trilobite treasures could be stumbled upon.

The more curious of these visitors to La Paz soon learned that for decades—if not centuries—industrious local inhabitants had been exploring the surrounding hillsides, looking for the hard, rust-hued concretions that had weathered out of the area's 10-meter-thick mudstone layers. While many of these discoveries would prove barren, occasionally they would yield Devonian fossils: sometimes a crinoid stem, perhaps a conularid or brachiopod, but predominantly disarticulated bits and pieces of the region's unique assemblage of trilobite species. Although almost always unidentified at the source, these usually less than perfect examples would subsequently reveal themselves to include such unusual trilo-types as *Dipleura boliviensis*, *Viaphacops orourensis*, *Eldredgia venustus*, and *Belenopyge balliviani*.

In a particularly poor region of a generally impoverished nation (where the annual per capita income late in the twentieth century was less than $1,000), these locals quickly discovered that there was a true hunger for their fossil finds . . . as well as a good chance for money to be made. They may not have particularly cared about either the

to change in an adventure-seeking post-World War II environment. At various times throughout the second half of the twentieth century, European and North American scientific expeditions put forth the extended effort required to explore the towering fossil-filled deposits that dot the La Paz area. The inaccessibility of the terrain, along with the energy and expense needed to assemble a paleontological team ready to sojourn to such a distant frontier, made any complete trilobite specimens gleaned from these ventures incredibly coveted commodities.

great age or the corresponding scientific gravitas of these appealing trilobites, but they certainly did care about their perceived financial value. And better yet for these determined Bolivian merchants, not only did teams of visiting scientists seek out these Paleozoic "trinkets" but so did tourists from around the globe.

By the 1970s, as many as a dozen resident trilobite hunters would traverse into the surrounding hills on a daily basis to forage through the territory's plentiful fossil exposures. Whether their concretion-covered discoveries were marketed to those who viewed such finds as true paleontological prizes or merely as native-generated curios, the economic welfare of La Paz's abundant Devonian formations was one of the local inhabitants' primary monetary resources until the mid-1980s, when the area's rampant drug trade started to transform the culture.

"On many levels, it was very important work," said Colon. "The trilobites these natives found were more than enough to start drawing serious scientific interest in the direction of Bolivia."

For some academics, the Devonian discoveries emanating out of the La Paz region have been a fossil-spawned bonanza. Indeed, judging by the quantity, quality, and diversity of the trilobites extracted from these South American sedimentary layers, it's easy to understand why, time and time again, these ancient sea creatures have drawn scientists from throughout the world to this distant spot on the now landlocked Bolivian landscape. Many of the trilobite species found in the area's mid-Paleozoic formations are indigenous only to these outcroppings, with their special status apparently due to a few key environmental factors, perhaps the most important being the geographical position in the Devonian held by what is now the soaring Altiplano.

During that distant period of Earth's history, what is now Bolivia was part of a rich ocean

PENNAIA VERNEUILI D´ORBIGNY, 1842
Middle Devonian; Sicasica Formation; La Paz, Bolivia; 5.1 cm

This is a particularly well-preserved example of this species. Often, the detailed compound eyes are left in the negative half of the split concretion.

ecosystem located just off the supercontinent known as Gondwana. For the first time in tens of millions of years, the offshore continental shelves that would eventually emerge as the western slopes of South America had become separated from similar shallow reef habitats to the north. Over time, this degree of pelagic isolation along with occasional dramatic shifts in climate provided the opportunity for new trilobite genera to emerge—a circumstance apparently seized upon with considerable relish by the area's resident arthropods. The Devonian also bore witness to a dramatic rise in sea levels, which followed a lengthy period of diminishing marine biozones during the tail end of the preceding Silurian. This served to radically alter certain long-established trilobite habitats while simultaneously creating fresh ecosystems within which divergent species could develop and thrive.

"When you first see the trilobites from Bolivia, you may not be that impressed," Colon said. "But when you start to examine them more closely, and note their individual qualities, you begin to understand that these trilobites come out of what may be some of the more important fossil deposits in the world."

The unique morphological characteristics exhibited by the trilobites found within the Sicasica and Belen, as well as their neighboring Devonian formations, have inspired their fair share of important scientific work. In the early 1970s, the renowned paleontologist Niles Eldredge, curator emeritus of New York's American Museum of Natural History, began extensive research on the area's phacopid and calmonid species. His published studies rocked the fossil world to its core, revealing previously unknown and unimagined details of trilobite anatomy and lifestyle as well as presenting fascinating insights into the evolutionary relationships between varied but occasionally closely related trilobite genera.

Indeed, Eldredge's widely hailed 1980 work on Bolivian arthropods, *Calmonid Trilobites of the Lower Devonian of Bolivia*, proved vital in supporting his and the late Stephen Jay Gould's 1972 theory of punctuated equilibria describing the fits-and-starts nature of evolutionary change. Many previous hypotheses—dating back to Darwin himself—had presented the notion that evolution was a slow and steady process, with change happening at more or less constant intervals throughout the life span of a given species. Punctuated equilibria put forth the rather radical concept that most species could exist in a state of relative calm, or stasis, for perhaps millions of years before various environmental pressures forced them to evolve over relatively brief periods of time.

"The Devonian faunas of the Southern Hemisphere have been studied sporadically since the mid-nineteenth century," Eldredge stated in his landmark effort. "But the basic descriptive analysis of the fauna was far from complete. Then in 1972, LeGrand Smith, then of La Paz, visited the American Museum of Natural History and donated trilobite specimens of what would be later described as *Legrandella lombardii*. Smith became a catalyst for much of the scientific work that followed."

Eldredge's groundbreaking studies of Bolivian trilobites served to focus additional academic attention on these remote sedimentary repositories. Yet late-twentieth-century travelers to La Paz—at least those with a primarily paleontological agenda—remained relatively few and far between. One of the more noteworthy visitors was the aforementioned LeGrand Smith, a minister from North Carolina who lived in the La Paz region in the late 1960s and early 1970s and did much to introduce the area's amazingly varied Devonian trilobite fauna to the world's scientific community.

Reaching any of Bolivia's trilobite-bearing formations continues to present a daunting, if not necessarily impossible, challenge for even the most determined scientist, collector, tourist, or adventurer. These difficulties stem from a combination of factors, most centering on the area's inaccessible location, as well as the still tangible dangers presented by local drug lords, few of whom could distinguish a trilobite from a tater tot, even if their

***METACRYPHAEUS GIGANTEUS* (ULRICH, 1892)**
Middle Devonian; Belen Formation; La Paz, Bolivia; 12.3 cm

This is a highly unusual specimen both for its size and for the fact that it was found in shale layers rather than within an encasing concretion.

lives depended on it. Despite inconsistent government rulings surrounding the collection and exportation of their nation's renowned natural resources, at certain times over the last half-century a steady flow of top-quality Bolivian trilobite material has managed to infiltrate the world's fossil marketplace.

In recent years, an upstart generation of native fossil diggers has begun to emerge in and around La Paz. Many members of this "new breed" seek previously untested avenues through which to advertise the unique concretion-encased fauna found within the area's Devonian formations. They have even begun utilizing the internet to present their latest discoveries to an increasingly receptive international clientele. Although their positive/negative trilobite specimens generally lack the often colorful calcite shells seen on more flamboyant arthropod examples hailing from other parts of the planet, these merchants have quickly learned that there is a palpable and profitable worldwide interest in their extraordinary fossil finds. They have subsequently requested and received limited government permission to market their Paleozoic product, and by doing so have helped reestablish Bolivia's reputation as a major—and reputable—outlet for trilobites.

Even a casual internet search often reveals exceptional, complete examples of such once-rare Bolivian trilobite species as *Wolfartaspis cornutus*, *Malvinella buddae*, and *Bouleia daginacourti* available for sale, usually at a reasonable price ($300 or less). But with more and more restrictions now being placed on digging for paleontological treasures throughout the La Paz vicinity (with some of the nation's long-standing and stringent archaeological laws being shoe-horned to carry over to the fossil realm), many of these attractive and important species may soon be in short supply, and those that do remain obtainable are almost sure to escalate in their cost of acquisition.

No matter what may happen regarding the availability of these South American trilobites in upcoming years, the insights they have provided into a long-gone Paleozoic kingdom have been of incalculable value to both collectors and academics around the globe. Even as recently as the turn of this century, many paleontologists might never have imagined that they would be presented with the opportunity to encounter such an incredible diversity and abundance of Bolivian trilobite material. These intriguing arthropods, drawn from the fossil-rich sedimentary strata of the sky-high Altiplano, are more than enough to take any trilobite enthusiast's breath away.

TRILOBITE SPINES

It seems safe to surmise that the life of a trilobite was filled with daunting challenges. From their first days in the Lower Cambrian to their last moments at the end of the Permian—more than a quarter of a billion years later—these primeval arthropods faced constant, increasingly advanced predatory threats. The trilobites' initial evolutionary response to such peril was perhaps the most

(OPPOSITE PAGE, TOP ABOVE) **HOPLOLICHOIDES CONICOTUBERCULATUS (NIESZKOWSKI, 1859)**
Upper Ordovician, early Caradocian; Kukruse Regional Stage; Viivikonna Formation; Alekseevka Quarry; St. Petersburg region, Russia; 6.5 cm

When spinose species such as this first began appearing on the world stage late in the twentieth century, most collectors and museum officials cast a wary eye in their direction. Since then, their authenticity has been firmly established.

(OPPOSITE PAGE, BOTTOM) **OLENOIDES SUPERBUS (WALCOTT, 1908)**
Middle Cambrian; Marjum Formation; House Range, Millard County, Utah, United States; 12.4 cm

With its large size and rows of freestanding spines, this specimen ranks as an incredible example of Paleozoic perfection.

important; throughout their long history, their thick calcite shell provided at least a degree of protection from attacks by sharp-jawed Cambrian sea monsters, 2-meter-long Silurian aquatic scorpions, and razor-toothed Devonian fish.

However, what happened next to the trilobite body design proved to be as revolutionary as it was evolutionary—the appearance of a complex, and apparently highly effective, series of spines that eventually emanated from every segment of some trilobites' dorsal anatomy. Exactly when trilobites started to generate such distinct external ornamentation is a question still open to scientific speculation and debate. A few early members of the trilobite line, such as the 520-million-year-old *Esmeraldina rowei*, did possess a rudimentary form of this adaptation. But by the Middle Cambrian, some 10 million years later, such species as *Kootenia randolphi* and *Olenoides superbus* clearly featured an imposing series of spines that both projected in a straight line down their axial lobe and encircled much of their thorax and pygidium.

It has been proposed that these primal spines—constructed from the same calcite-infused materials as their outer shells—may have been somewhat flexible, and perhaps were initially utilized by certain trilobites as a basic style of navigating rudder. Some academics believe that they may have served an important role in sexual display and courtship. Others have even speculated that these spines may have been attached to nerve-filled sensory organs, providing the trilobite with a Jedi-like ability to feel disturbances in the water around it. Still others postulate that these primitive spines represented

ESMERALDINA SP.
Lower Cambrian; Waucoban Series; Esmeraldina County, Nevada, United States; 9 cm

This 520-million-year-old species may be among the first trilobites in the fossil record to display spines along its axial lobe.

nothing more than the start of an undersea "arms race" that saw creatures around the globe gearing up for daily battles of survival.

In all honesty, it's not as if paleontologists had long assumed that these arthropods possessed rows of pronounced defensive spines. During the first few centuries of trilobite research, dating back as far as the mid-1800s, such significant yet fragile morphological features were virtually unknown and usually were not recognized in fossilized trilobite remains. It wasn't until more refined preparation techniques began to flourish in the 1980s that the frequently outrageous array of quills, barbs, spikes, and nodes that adorn certain trilobite carapaces began to be uncovered in all their spinose glory.

It didn't take long for either the collecting or the scientific communities to take notice of these extraordinary trilobite species. The worldwide commercialization of both the diverse Devonian trilobite fauna of Morocco and the beautifully preserved Ordovician biota hailing from Russia helped launch a serious reconsideration of the role spines may have played in the lifestyles of these pervasive Paleozoic creatures. During the last decades of the twentieth century, an almost overwhelming assortment of freakishly spiked trilobite exoskeletons started to emerge from these locales, bewitching—and at times bewildering—even the most perceptive members of the trilobite community. Magnificently prepared examples of *Quadrops flexuosa*, *Comura bultyncki*, and *Hoplolichas tricuspidatus*—each featuring a unique, varied, and menacing pattern of pointed spikes and spurs—served to dramatically showcase the varied styles and shapes that trilobite spinosity exhibited in those primitive seas.

This proliferation of larger and more intimidating trilobite spines apparently reached its apex in the Devonian. At that time in Earth's history, Moroccan species such as *Drotops armatus*, which routinely reached lengths of 15 centimeters, had evolved into creatures resembling heavily armored Paleozoic battle tanks—formidable invertebrates covered in dozens of centimeter-long spikes. And judging by the impressive span of time during which many of these spine-encrusted species survived, it would certainly appear that such a defensive/mating/sensing adaptation served its purpose quite effectively. Indeed, it is now indisputable that the evolutionary development of spines played a significant role in trilobites being among the most successful creatures to ever grace the face of our planet.

HARAGAN FORMATION, OKLAHOMA: *ONE MAN'S PASSION*

Near the southern border of Oklahoma, tucked neatly within the surrounding Arbuckle Mountains, sits Coal County, one of 77 districts that checkerboard the Sooner State. From the local hamlet of Clarita, it's less than a two-hour drive north to Oklahoma City, and about a three-hour trek south to Dallas. But in terms of cultural outlook and regional attitude, such sojourns may as well be measured in light years. Fewer than 6,000 people live in the county's 840-square-kilometer parameters, and those that do reside within this dusty stretch of land tend to both accept and revel in their decidedly blue-collar standing. At a very different moment in U.S. history, the region that now comprises Coal County was part of the Choctaw Nation, located in the heart of what was then officially known as Indian Territory. Obviously, much has changed about the area and its demographics since the nineteenth century, although certain elements of Coal County seem perpetually mired in the mists of time.

As one might surmise from its name, coal mining has played a significant role in this county's story dating all the way back to when the Native

DICRANURUS HAMATUS ELEGANTUS (CAMPBELL, 1977)

Lower Devonian; Haragan Formation; Coal County, Oklahoma, United States; 10.1 cm

This trilobite species is remarkably similar to a Moroccan "cousin" in both appearance and lineage. At this time in Earth's history, the land that would eventually become Morocco and Oklahoma shared a contiguous marine environment.

American population still had a major say about the way things were done in and around these parts. The pinnacle of coal production lasted for a relatively short period, basically from 1870 until 1920, after which all the mines began to disappear. Since then, things have often been economically challenging in this section of the state, with even the county's stabs at agriculture being limited by occasional boll weevil infestations. Just about any way you look at it, these days nothing particularly exciting seems to happen within this remote parcel of Oklahoma real estate. Sure, there's plenty of hunting and fishing, and even an annual Amish Crafts & Antiques Show. But despite its notable lack of high-gloss affectation, don't be deceived by the backwater vibe the region so effortlessly exudes.

If you happen to be interested in trilobites, this often arid, rough-hewn tract of land may well be the perfect place to make your heart beat just a little bit faster. You see, much of Coal County sits on rich Lower Devonian limestone layers filled with some of the best-preserved trilobites in the world. One of the fortuitous by-products of the extensive coal mining that took place more than a century ago was the discovery of a variety of promising fossil-bearing localities. These featured everything from ice-age mammoth bones, to Triassic dinosaur deposits, to the vicinity's impressive array of Silurian and Devonian trilobite outcrops. Few of these sites were properly explored at the time, but by the middle of the twentieth century, a small number of intrepid adventurers led by local geologist J. W. Stovall began to more fully investigate Oklahoma's fossiliferous holdings.

After satisfying their initial dinosaur-centric predilections, these scientists turned their attentions toward the state's Paleozoic horizons. What they almost immediately discovered was that the 417-million-year-old Haragan and the closely aligned Bois d'Arc formations were both heavily represented throughout the region. They also noted that these outcroppings offered a wonderfully detailed look at diverse Devonian fauna that presented compelling evidence of a once thriving offshore community. These rich sedimentary layers were brimming with bryozoans, bivalves, and brachiopods as well as trilobites, all perfectly preserved within a finely grained limestone matrix.

Slowly but surely, awareness of this fascinating mid-Paleozoic material started to filter through the paleontological community. By the 1960s, scientists working at the University of Oklahoma's

***DOLICHOHARPES RETICULATUS* WHITTINGTON, 1949**

Upper Ordovician; Bromide Formation, Pooleville Member; Criner Hills, Oklahoma, United States; 2.2 cm

This attractive and unusual trilobite is found on rare occasions within the Bromide. A nearly identical species exists within the Platteville Formation of the American Midwest.

Sam Noble Museum had launched their own preliminary investigation into the fossil-filled Coal County outcrops. Their efforts revealed enticing but generally fragmentary evidence of such notable trilobite species as *Dicranurus hamatus elegantus, Huntonia lingulifer, Ketternaspis williamsi,* and *Reedops deckeri,* among many others. Despite these initially promising results, it would be another 20 years before the proper degree of attention began to again be paid to the trilobites of the Sooner State.

The most significant figure behind this renewed interest in the area's abundant trilobite resources has proven to be a somewhat eccentric character by the name of Bob Carroll. With his shaggy mop of brown hair, scruffy graying beard, look right through you stare, and occasionally gruff demeanor—especially around those he doesn't know—Carroll cuts an imposing image as the kind of guy who doesn't deal particularly well with those he perceives as wasting either his time or his energy. What you see with Carroll is exactly what you get, a Vietnam-era Navy veteran who clearly prefers to work alone than be surrounded by a crowd; someone who'd rather be playing some nasty blues licks on his left-handed electric guitar than discussing the nation's latest political twists and turns.

For the last three-plus decades, Carroll has diligently and single-handedly worked the Devonian formations in Coal County, and he has done so with a dedication and focus that few others in the fossil field can match. By the time he reached Oklahoma in the mid-1980s, he had already enjoyed a long history exploring for trilobites, spending nearly a decade prospecting throughout the Midwest. Once Carroll learned of the diversity and beauty of the trilobite material available in this off the beaten track southwestern station from a fellow trilobite enthusiast named Fred Wessman, he realized there was no turning back. His work in the sedimentary exposures of Coal County quickly became more than a job; it became his life's mission.

Each year Carroll spends countless hours from late March until early October laboring outside in the harsh elements that too often sweep through the 20-odd acres of leased quarry land that surround his home near Clarita. There, on what has become known to virtually everyone in the fossil community as Black Cat Mountain, he carefully breaks apart layer upon layer of the Haragan Formation's characteristic pale yellow matrix in search of the caramel-colored calcite remnants of his often elusive Paleozoic prizes. After finding a requisite number of cross-sections of what he

hopes will be complete trilobites, Carroll retires to his on-location prep lab and gazes through the lenses of his high-powered binocular microscope for hours on end, artistically transforming the best of his fossil finds into some of the most intricate and desirable trilobites ever found anywhere.

"People see the finished product of my work, but they usually don't even consider the amount of time and effort that goes into finding and then cleaning even the smallest trilobite," Carroll said. "I love what I do, but it can also be very frustrating. I'm a perfectionist when it comes to my trilobites, so sometimes I think I'm working on what will turn out to be a great specimen, and then when I'm about 90 percent done, I'll find that a genal spine will be missing, or the pygidium will be slightly disarticulated. But even the ones that I think aren't perfect, people really do seem to like."

When engaged in conversation, Carroll often bemoans how the summer's heat limited his time in the field that season, how the shifting whims of the economy have negatively affected his sales, or how "bad luck" stopped him from finding enough marketable examples of his favorite trilobite species. When properly prodded, he'll rather casually mention how he's "had enough" with the unpredictable nature of fossil collecting . . . and fossil collectors.

Despite the aura of doom and gloom that he often projects, each year in late January, when you walk by his showroom on the first morning of the fossil mega-show held in Tucson, Arizona, you'll find that enthusiasts from around the globe have already lined up, jostling with one another in their attempts to be the first to see—and probably purchase—Carroll's latest trilobite masterpieces. For all the frenzied aura that often abounds in Carroll's showroom, just about the last thing this Michigan native believes in is hyping his own product. Much like the man himself, either you

ISOTELUS SP.

Ordovician; Viola Springs Formation; Pontotoc County, Oklahoma, United States; 14.2 cm

This specimen is one of the more unusual trilobites found within this rarely explored formation. With only seven exposed thoracic segments, this is probably a molted exoskeleton.

like Carroll's work, or you don't. And, to be honest, he doesn't seem to particularly care about your reaction—one way or another.

"Dealing with Bob is always an interesting experience," said Sam Stubbs, a lawyer from Houston who owns one of the finest collections of Carroll-generated Oklahoma trilobites. "He possesses so much knowledge, and the quality of his work is so good, that you just want to draw as much information out of him as you can. However, he's not always that willing to reveal too many of his secrets."

There's good reason that over the last four decades the trilobites emerging from Carroll's small prep lab have created a veritable feeding frenzy among those who collect and study these intriguing invertebrates. With their lifelike preservation, distinctive golden color, free-flowing spines, and intricate surface textures, these ancient marine relics are beautiful enough to serve as the star attractions in just about any major museum exhibit or private trilobite display. In addition, the scientific information gleaned from Carroll's ongoing work has been more than sufficient to amaze paleontologists from Tulsa to Tokyo. Prior to Carroll's efforts, rarely—if ever—were complete Oklahoma trilobites featuring the bizarre morphology seen in the "ram-horned" *Dicranurus hamatus elegantus*, the big-eyed *Viaphacops bombifrons*, and the spinose *Ceratonurus sp.* available for academics to hold, appreciate, and examine.

Even those who garnered their doctoral degrees based on the study of partial, surface-collected specimens will grudgingly admit that incomplete material can only offer tantalizing hints regarding a trilobite's true tale. Once Carroll began full-scale operations in the heart of Coal County in 1986, it became apparent that his quarry would produce fully intact examples of an amazing assemblage of seldom before seen trilobite species. It also became evident that these trilobites would be relatively abundant and, perhaps most important, that most would be of unique postpreparation quality.

Carroll won't even hazard a guess as to how many examples of each of the more than 20 trilobite

(OVERLEAF) **VIAPHACOPS BOMBIFRONS (HALL, 1861)**

Lower Devonian; Bois D'Arc Formation, Cravatt and Fittstown Members; Coal County, Oklahoma, United States; 6.3 cm

This is a large, nearly perfect example of one of the rarer phacopid species to be found within the Coal County environs.

species found in the Haragan and Bois d'Arc formations he's discovered and prepared over the last 35 years. When further nudged on the subject, he surmises that he's "probably found thousands" of the quarry's most common trilobite, the small phacopid *Kainops raymondi.* And he admits that he's prepped more than a hundred complete examples of the two types of *Huntonia* (recently changed to the less trippingly on the tongue *Huntoniatonia* to avoid taxonomic confusion with a genus of isopod) that inhabit Black Cat Mountain.

This numbers game gets a lot easier for Carroll once he begins to focus his attention on the quarry's truly rare species. There have been perhaps two dozen complete examples found of the diminutive lichid *Acanthopyge consanguinea,* and only a smattering of the 4-centimeter ovate *Breviscutellum oklahomae.* He has found but a single complete specimen of the spectacular 3-centimeter-long spiny odontopleurid, *Laethoprusia sp.,* and the discovery of that trilobite, now housed in a private Japanese collection, remains one of the highlights of Carroll's long career.

"I'll remember the day I came across that *Laethoprusia* for the rest of my life," he said with a smile. "It's kind of small, so there was a good chance that I could have missed it. I'm mighty glad I didn't!"

Somewhat ironically, when Carroll was beginning operations at Black Cat Mountain in the mid-1980s, equally rich Devonian trilobite reserves of almost the exact same geologic age were being discovered more than 8,000 kilometers away in the Atlas Mountains of Morocco. Over 400 million years ago, what is now North Africa was located off the coast of the landmass known as Gondwana, and present-day Oklahoma was situated in shallow seas adjacent to the continent of Euramerica. Both trilobite-filled sanctuaries then existed in relatively close geographic proximity, on similar equatorial latitudes, and presented almost identical marine environments. Thus, the fact that many nearly indistinguishable trilobite genera (including *Dicranurus, Ceratonurus, Reedops,* and *Breviscutellum*) lived and apparently thrived in these dual Paleozoic outposts would initially seem far from surprising. Yet the remarkable similarities between the trilobite fauna found in these two now disparate localities—although the Moroccan material is generally preserved in a thick, charcoal-hued calcite—continues to amaze even the most seasoned collectors. This Devonian confluence also serves as a comfort to academics who believe that such rock-solid fossil evidence confirms their confidence in the foundational geological concept of plate tectonics—the theory that explains how large continental land masses ever so slowly drifted across the Earth's outer crust over extended periods of time, carrying their embedded fossils with them.

Carroll finds such scientific discourse interesting, but he admits it's all far removed from what motivates him to venture into the field each morning. Come rain or shine, heat or chill, hell or high water, he'll be out there atop Black Cat Mountain breaking rock and keeping a close watch for the telltale signs of rare trilobites, as well as for the area's

---

(OVERLEAF) *BREVISCUTELLUM SP.* **CAMPBELL, 1977**
Lower Devonian; Bois d'Arc Formation, Fittstown Member; Coal County, Oklahoma, United States; 4.6 cm

Only a few complete examples of this species have been found in the Bois d'Arc, a formation closely aligned with the neighboring Haragan. Each formation does, however, present a markedly different trilobite fauna.

(OPPOSITE PAGE) *HUNTONIATONIA HUNTONENSIS* **(ULRICH AND DELO, 1940)**
Lower Devonian; Haragan Formation; Black Cat Mountain Quarry; Clarita, Coal County, Oklahoma, United States; complete specimen: 11.3 cm

This unusual "double" was the highlight of the 2018 collecting season at Black Cat Mountain.

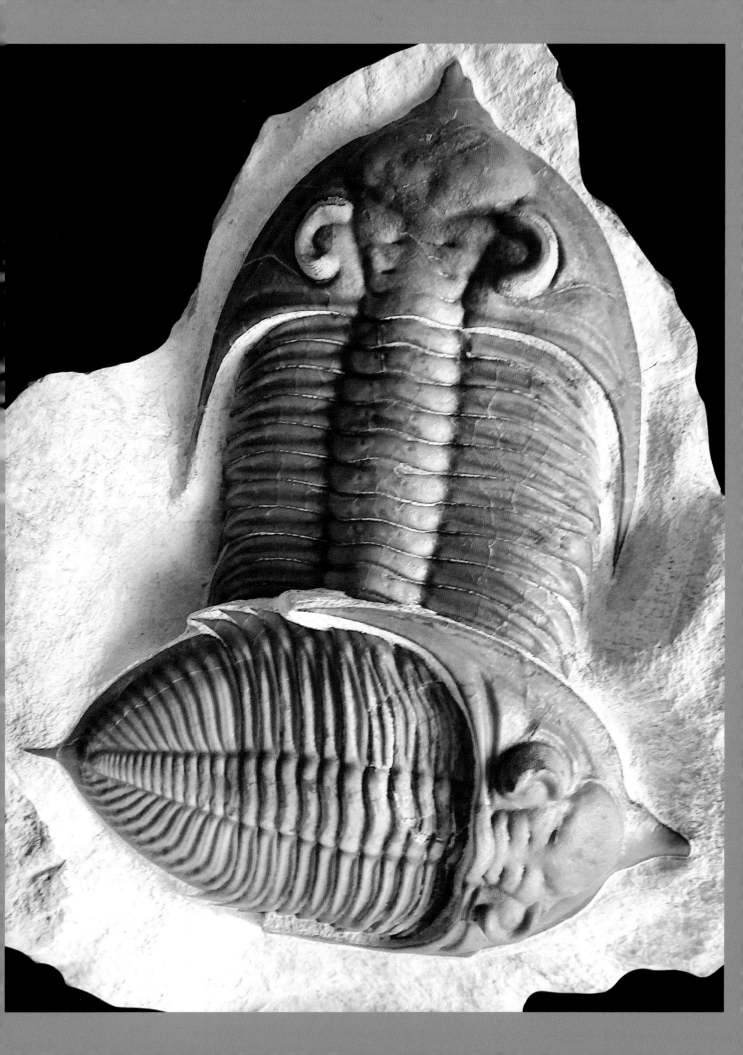

notorious rattlesnakes. In the eyes of his many admirers, Carroll is more artist than academic, one of the few fossil preparators who has taken the job of cleaning trilobites to new levels of virtuosity and craftsmanship. He, perhaps more than anyone else, has helped put the tiny town of Clarita, Oklahoma, on any map featuring the world's most significant trilobite localities. Of course, true to his gruff yet unassuming nature, Carroll will quickly shrug off such praise with an off-handed "thanks" before adding, "I'm just doing the best I can. Hey, this is what I do for a living."

TRILOBITE MOLTING BEHAVIOR

Considering the endless series of evolutionary obstacles they faced during their lengthy stint in the planet's primeval seas, it's sometimes hard to believe that trilobites endured in such a hostile environment for more than 270 million years. Despite their lives being in constant peril, perhaps their moments of greatest vulnerability didn't occur when they encountered razor-toothed marine predators, undersea methane eruptions, or mountainous mud slides. Perhaps those times came when they shed their hard outer shells and, by necessity, revealed their "softer" side to the world. It was then, during periods of molting—when they jettisoned those old calcite carapaces and began growing larger, thicker new ones—that trilobites were left virtually defenseless against the multitude of competitors that continually battled them for control of their aquatic domain.

Some scientists have postulated that trilobites may have been the first organisms ever to utilize the molting process to expedite and facilitate their growth, a somewhat Sherlockian supposition when one considers that these arthropods were also among the first animals in our planet's history to possess a protective shell. And the fact that some trilobites attained prodigious sizes—on

rare occasions reaching 70 centimeters in length—suggests that molting behavior may well have continued throughout their potentially prolonged lives, although in modern arthropods such as lobsters molting slows down considerably as the animal ages and grows. (The largest lobster on record, found in waters off Nova Scotia in 1977, was 106 centimeters long, weighed in at a hefty 20 kilograms, and was estimated to be at least 100 years old!)

The fossil record also indicates that molting emerged right along with the earliest trilobites, more than 520 million years ago. By the time the first trilobite species started filling Lower Cambrian seas, they already possessed those uniquely calcified exoskeletons, and soon after (by the Middle Cambrian) they would develop clearly delineated cephalic suture lines that provided them with the ability to more effectively shed their shells during each successive growth cycle. Thus, the overwhelming majority of recovered trilobite fossils are of this molted, often disarticulated outer armor rather than of the actual animals themselves.

Indeed, it is believed that many of the impressive trilobite mass-mortality plates retrieved from key fossil-bearing locations around the globe (including *Homotelus* from Oklahoma, *Xenasaphus* from Russia, and *Eldredgeops* from New York) may be the resultant by-product of a specific trilobite species gathering together during times of molting. In retrospect, it makes perfect sense for a community to congregate at periods of peak vulnerability and peril for mutual protection. To execute the molting process, most trilobite species needed only to

BATHYNOTUS KUEICHOWENSIS (LU, IN WANG ET AL., 1968)

Cambrian Series 2, Stage 3; Kaili Formation; Jianshan Section; Jianghe County, Guizhou Province, China; 7.3 cm

The molted glabella of this large trilobite lies to the right of its thorax. Until recently, such specimens were known almost exclusively through line drawings in decades old Chinese text books.

anchor their tail ends on the surrounding seafloor, flex their bodies, and pop out through the easily detachable sutures on either side of their heads. Hmmm, it all sounds so easy . . . so efficient . . . so evolutionary. And it probably was—at least for a while.

A recent scientific paper by Michigan State University paleontologist Danita Brandt indicates that their increasingly antiquated molting behavior may have contributed to the slow yet steady decline of the trilobite line. This report states that as time marched inexorably on, the improvisational method used by these arthropods to shed their shells, as well as their apparent inability to reabsorb (through ingestion) the mineral-rich resource provided by their discarded exoskeletons, proved highly inefficient—and potentially devastating. Although such apparent morphological disadvantages were probably not the primary cause of the trilobites' eventual disappearance, they may have played a significant role in their gradual but continuous downturn throughout the Paleozoic.

Of course, any organism that existed for as long as the trilobite may have naturally outlived many of its initial anatomical advantages. If their molting behavior eventually became one of those increasingly obsolete characteristics, all I can say is that the trilobite line still enjoyed one "shell" of a ride through early Earth's history.

SILICA SHALE, OHIO: *TRILOBITES FROM THE HEARTLAND*

As just about anyone who has ever searched for a trilobite already knows, the fossilized remains of these Paleozoic arthropods are often found in the most out-of-the-way places: burning deserts, frigid mountain tops, treacherous seaside cliffs, inaccessible river valleys—and in old cement quarries. This last entry may bear little of the majestic *Nat Geo* aura of the others, but without a doubt the

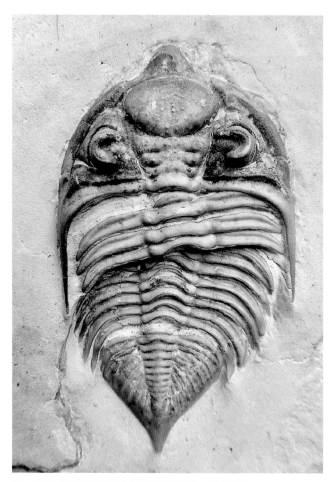

DALMANITES HALLI (WELLER, 1907)
Silurian; Waldron Shale Formation; Shelby County, Indiana, United States; 5.2 cm

This is a slightly disarticulated but visually appealing example of a rare trilobite from the famous Waldron outcrop of the American Midwest.

most famous cement quarry in the paleontological world is stationed in Sylvania, Ohio—approximately 25 kilometers west of beautiful, downtown Toledo.

For more than a century, on land exclusively controlled by the Medusa Stone Company, tons of what was once a thriving Devonian-age seabed have been commercially extracted for use as road fill, driveway gravel, and, yes, a basic ingredient in cement. Due to the frenetic men at work nature of quarry operations, as well as the dangerous rock-moving machinery that on occasion still lurks in the vicinity, outsider access to this outcrop of Silica Shale has always been severely limited. However, if you've ever been lucky enough to gain entry to the quarry's tree-encircled compound—whether by invitation or in a somewhat clandestine weekend raid—no matter where you look it is almost impossible not to encounter evidence of life-forms that lived and apparently flourished here nearly 400 million years ago, in what was then the bottom of a shallow, semitropical bay.

Mixed in with the coarse, light gray sedimentary mudstone that comprises the quarry's primary matrix is a vast assortment of fossiliferous material: brachiopods, coral crowns, and crinoid stems. In certain places these intricately detailed remnants are so pervasive that they seem to dominate the silica-rich rock in which they've been trapped for nearly half-a-billion years. It is hard to move more than a few feet anywhere within the site's sizable boundaries without spotting a Paleozoic fossil of some kind. But most of the collectors who wander into the Sylvania locale aren't particularly interested in the abundant array of brachs and crinoids they may encounter, although some enthusiasts do specialize in collecting, studying, and even selling such material that, after a little prep work, often displays an attractive pyritized appearance. To put it simply, virtually everyone who visits the quarry is there to find one thing—trilobites.

Nearly all of the quarry's trilobites belong to two strikingly similar phacopid species: *Eldredgeops milleri* and its nearly identical cousin *Eldredgeops crassituburculata.* Beautifully preserved in either a charcoal gray or chocolate brown calcite, these incredible three-dimensional arthropods—which can be found both tightly enrolled in a defensive position or totally outstretched and usually range between 4 and 7 centimeters in length—appear little the worse for wear despite their lengthy passage through evolutionary time. Each of the facets adorning their extraordinary compound eyes

are easily observed, with there being a slight, yet distinguishable difference in alignment, size, and shape between the lenses of the two *Eldredgeops* types. For many collectors, this nominal variance serves as the primary means of both identifying and separating these captivatingly comparable species.

Yet there is more to the appeal of these 390-million-year-old Devonian arthropods than just their astounding, honeycomb-like ocular attachments. Sylvania's trilobites (the site also features a rare proetid, *Basidechenella lucasensis*) are celebrated for revealing the dorsal structure of their carapaces in exquisite detail. The unmatched quality and quantity of these fossilized organisms have helped elevate this Silica Shale outcrop to a lofty Paleozoic plateau—one that places it atop any list featuring the planet's best-known Devonian exposures. Similarly, collectors who have been fortunate enough to explore this bountiful paleontological resource over the decades have long since established themselves as among the most often envied inhabitants of the trilobite community.

Those who have traditionally collected fossils within the Sylvania quarry have earned the nickname Silica Brats, a derisive-sounding term that those targeted proudly wear as a badge of honor. They know that being recognized in this way carries a certain degree of prestige among their fellow trilobite enthusiasts—those who acknowledge the continual dedication these "brats" have exhibited toward gathering the Silica Shale's diverse Devonian-age fauna as well as their luck for living within the parameters of Ohio's Lucas County, making it a relatively easy drive to reach this renowned paleontological repository.

Over the years, scores of trilobite collectors have grown up in and around the distinctly suburban environs that surround this fabled, fossil-filled quarry. Most fondly recall the idle hours they spent wandering through the site's slag heaps and

rock piles, searching for prized Paleozoic relics. As they grew older and life's inherent responsibilities began eating up much of the carefree time of their youth, the passion these midwestern hobbyists have maintained for this Silica Shale deposit—and its trilobites—has been nothing short of astonishing. Many of these intrepid collectors still venture into the quarry every chance they get, whether it's on a long summer Sunday or a random spring afternoon. Most do so strictly to satisfy their never-ending quest to find the ultimate *Eldredgeops* specimen—that elusive big-eyed, perfectly aligned, fully inflated, totally outstretched 10-centimeter-long "monster" of their dreams.

At the slightest provocation, these ardent arthropod aficionados will happily regale you with "trilobite tales" about the flawless specimen their digging buddy found last year working just a step away from them, or how the "biggest bug I've ever seen" was broken by a slightly overzealous hammer blow, or how a legendary trilobite may finally be freed from someone's private collection within the next few months. It's apparently a life-long game among the Silica Brats to boast about who has the biggest, the best, the most perfect Sylvania trilobite, although stories of humongous "lost" specimens and covert missions into closed quarry sections also serve as frequent fuel for after-dinner discussions.

It breaks the collective heart of these Silica Shale devotees to consider the fate suffered by a preponderance of the quarry's fauna. The sad truth is that for every trilobite found in the area's rich outcrops throughout the last dozen decades—and there have literally been thousands in a variety of sizes, shapes, and states of preservation—probably 10 times as many have been chewed up and spat out by the huge cranes, backhoes, and compactors that have continually uprooted the fossil-rich layers of sedimentary rock. The resulting scree is then delivered into machines that crush it all, trilobites included, into gravel that will eventually fill roadbeds and driveways throughout the Midwest. It's enough to bring a tear to the eye of any serious trilobite enthusiast.

"I've been very lucky to have gone digging for trilobites in many parts of the country," said Ray Meyer, a noted trilobite authority and long-time Silica Shale aficionado. "But few locations have ever supplied more excitement, and kept me enthralled for longer than Sylvania. And the trilobites are truly amazing. They seem to have a soul . . . especially with those incredible eyes."

It is the considered although perhaps slightly biased opinion of many leading North American trilobite connoisseurs that no eyes in the fossil universe can match those found attached to Silica Shale's phacopid specimens. Proudly ensconced atop the creature's semicircular cephalon, these eyes hold an almost mystical fascination for those who gaze upon them. Indeed, their perfect geometric design and seemingly timeless symmetry appear to defy their incredibly primal pedigree.

The optically oriented details surrounding Sylvania's two phacopid species—especially the differing number of lenses adorning each eye's ventral file—have also helped scientists (most notably Niles Eldredge, curator emeritus of New York's American Museum of Natural History, for whom the area's *Eldredgeops* genus has been named) solidify their most prominent academic theories, including punctuated equilibria, which describes certain species as quickly mutating and evolving after lengthy periods of relative stasis. Whether they serve as a foundation for scientific

ELDREDGEOPS MILLERI (STEWART, 1927)

Middle Devonian (Givetian); Silica Shale Formation; Milan, Michigan, United States; largest trilobite: 7 cm

This outstanding cluster of legendary midwestern phacopid trilobites is not from the "classic" Ohio site but from a neighboring outcrop in Michigan, some 50 kilometers to the north.

research or as an inspiration for amateur collectors from Toledo to Taipei, few morphological features in the entire fossil record are as continually compelling—or as evolutionarily significant—as trilobite eyes.

By the time members of the trilobite class first appeared in the world's oceans some 521 million years ago, these arthropods already featured highly developed, mineralized optics. This monumental development marked them as the first creatures in the history of Earth capable of leaving behind fossil evidence of such a pronounced anatomical advance. Those original trilobite eyes were crescent-shaped, providing nearly 360-degree vision for the primitive members of the Redlichiida order that exhibited them—a major help when it came to keeping an eye (or two) peeled for the menacing cast of marine predators that surrounded them in those Lower Cambrian waters. Some 130 million years later, in the Middle Devonian, phacopid eyes had evolved into perhaps the most complex optical attachments *ever* to see or be seen. These large schizochroal eyes, featuring hundreds of individual lenses stacked in tight symmetrical rows, furnished its host trilobite with a truly remarkable view of the undersea world around it.

Unlike any modern eyes—whether arthropod, anthropoid, or annelid—all trilobite eyes were constructed of calcite, bestowing these ancient creatures with a unique and seemingly unparalleled ability to perceive changes in their marine environment. Evidence revealed by recent experiments conducted with calcite crystals suggests

ELDREDGEOPS CRASSITUBERCULATA (STUMM, 1953)

Middle Devonian (Givetian); Silica Shale Formation; Sylvania, Lucas County, Ohio, United States; 5.9 cm

Specimens of this Silica Shale trilobite occasionally grew to 9 centimeters in length. Rumors of larger examples run rampant but remain unsubstantiated.

that trilobite vision was filled with bold streams of light and vivid bursts of color—a veritable Paleozoic acid trip! Scientists continue to be fascinated by the evolutionary logistics behind the development of these amazing, mineralized ocular outlets, but trilobite collectors prefer to revel more in their singularly distinctive beauty.

"If there's one thing that the *Eldredgeops* from Sylvania are famous for, it's those eyes," Meyer said. "Of course, the overall preservation of each trilobite is also important, but a complete, outstretched specimen with perfect eyes is what everyone is looking for when they visit the quarry."

Reports of fossils being found in the Sylvania area have existed ever since the first settlers began arriving in this part of northern Ohio during the early 1800s. Back then, the locals were far more concerned with battling against the harsh winter elements and the occasionally inhospitable members of the native Shawnee tribe than with the possible paleontological implications of the shells made of stone (brachiopods) they frequently encountered while digging in their fields. Stories of strange fossil finds would periodically make local newspaper headlines for much of the century, but it wasn't until 1892, when the Sandusky Cement Company bought a tract of land in Sylvania and began quarry operations, that people started to pay greater attention to the strange, often inexplicable shapes that inhabited the vicinity's Devonian-age rock formations.

By the time the facility was renamed the Medusa Stone Company in 1929, the area's trilobites were already attracting attention from a variety of local museums and collectors. During the peak period for quarry operations in the mid-1960s, over 100 men were employed full-time at the site. Their contact with both Sylvania's fossils and the town's fossil collectors was constant—and to some extent problematic. Area trilobite enthusiasts continually pestered quarry foremen for the opportunity to

sneak into the facility and wade through the piles of enticingly extracted rock. These rabid collectors also frequently confronted workers with fervent requests to put aside any prime specimens for subsequent purchase or trade.

It is said that many great trilobites were obtained through these aggressive appeals, although these distractions were continually frowned upon by those running the quarry. They believed that "nothing good" could come from their workers associating with the county's passionate fossil collectors. However, these intriguing interactions between local trilobite devotees and Medusa employees continued virtually unabated until full-time quarry operations were suspended in 1979. Since then work at the Sylvania site has notably slowed, but the focus of the Silica Brats has not. Tales of giant 8-centimeter-long phacopid trilobites that have been secretly acquired in the dead of night continue to highlight certain local conversations.

"Dealing with the people at the quarry has always been a tricky subject," Meyer explained. "Some of the folks there were very cooperative with local collectors. They'd either look the other way, or they'd save special trilobites from being ground up in the machinery. But obviously, some of their bosses didn't look favorably on such activities. Their main concern was about someone breaking in without permission and getting hurt while looking for fossils."

The bosses' concerns were apparently merited. Over the years, any number of reported (and unreported) injuries have been incurred by somewhat rambunctious trilobite hounds rummaging around within the supposedly closed confines of the Medusa quarry. Despite the potential hazards they may have to face, it seems that every Silica Brat was, is, and always will be more than ready to pick up a trusty rock hammer at a moment's notice and venture onto the sacred soils of the Sylvania site. They each seem to know deep in their fossil-loving hearts that the biggest, best, most perfect *Eldredgeops* trilobite ever seen is right out there waiting to be found. And they're each determined to be the one to find it.

TRILOBITE EYES

As initially preposterous as it may seem, a strong case can be made that arthropod eyes rank as the single most important anatomical feature preserved in the entire fossil record. By the time the initial members of the trilobite class began appearing in Cambrian seas some 521 million years ago, they already possessed highly developed, mineralized eyes—marking them as the first creatures in Earth's then 4-billion-year history capable of leaving behind tangible evidence of such a major evolutionary advance. Current academic thought—based almost solely on finds recently made in the Lower Cambrian strata of Emu Bay, Australia—indicates that arthropod eyes may have evolved as much as 10 million years earlier than the trilobites' distinctive calcite exoskeleton.

Those original trilobite eyes were crescent-shaped and seemingly provided a nearly 360-degree field of vision for such primitive species as *Olenellus roddyi* and *Redlichia chinensis*. At approximately the same time, the carnivorous arthropod *Anomalocaris* had developed large-lensed compound eyes through which to glimpse its surrounding marine environment. Indeed, it has been speculated that the evolution of these highly functional eyes, and that advancement's

ELDREDGEOPS MILLERI (STEWART, 1927); ARTHROACANTHA CARPENTERI (CRINOIDS)

Middle Devonian (Givetian); Silica Shale Formation; Sylvania, Lucas County, Ohio, United States; 7.3 cm

This intriguing combination provides a captured-in-time moment along the Devonian seafloor.

DECHENELLA LUCASENSIS (STUMM, 1965)

Middle Devonian (Givetian); Silica Shale Formation; Sylvania, Lucas County, Ohio, United States; 3 cm

This smallish species is by far the rarest member of the Silica Shale trilobite fauna. It is estimated that as many as 500 complete *Eldredgeops* are found for every *Dechenella*.

subsequent impact on both predatory and prey animals within their primal marine ecosystem, may have helped kick the famed Cambrian Explosion into high gear.

By the Ordovician, approximately 50 million years later, trilobite eyes had developed into an almost bewildering assortment of sizes and shapes. Some, like those attached to such Russian species as *Asaphus kowalewski* and *Cybele panderi*, sat atop spindly stalks up to 5 centimeters in length. This highly specialized adaptation presumably allowed the trilobite a better view of the world around it, even when it may have been lurking under a thick layer of silt along the sea bottom. Other trilobites of the period, such as the

aptly named *Cyclopyge bohemica* and the closely related *Pricyclopyge binodosa*, displayed a single, huge, wraparound holochroal eye featuring hundreds of small, tightly packed lenses. This design allowed these free-floating species an uninterrupted look upon the ever-changing ocean floor as they carefully navigated their way through the primal seas.

Almost 100 million years later, in the Middle Devonian, trilobite eyes had evolved into the most complex ocular attachments ever to be found fossilized in the planet's sedimentary stratum. In some cases, these impressive eyes sat atop the trilobite's gently curving cephalon in a fashion most resembling the headlights of some fancy twenty-first-century European sports car. Phacopid species such as *Drotops megalomanicus* and *Eldredgeops milleri* had huge, schizochroal eyes featuring dozens of geometric lenses stacked in rigid, uniform rows—a pattern that allowed its

(OPPOSITE PAGE, TOP LEFT) **FENESTRASPIS AMAUTA (BRANISA AND VANEK, 1973)**

Lower-Middle Devonian; Upper quarter of the Lower Belén Formation; Chacoma, Bolivia; eye: 3 cm

The eyes that adorn this South American species may well be the largest of any trilobite that ever existed.

(OPPOSITE PAGE, MIDDLE RIGHT) **STRUVEASPIS BIGNONI CORBACHO, 2014**

Middle Devonian, Eifelian; Bou Tchrafine Formation; Jorf, near Erfoud, Morocco; 4.1 cm

The lens count on this trilobite's eye had dropped to a precious few, possibly indicating a lifestyle in turbid water where the need for keen eyesight had been minimized.

(OPPOSITE PAGE, BOTTOM LEFT) **SCABRISCUTELLUM FURCIFERUM (HAWLE AND CORDA, 1847)**

Middle Devonian; Bordj, Morocco; 5 cm

This trilobite species' holochroal eyes are covered in hundreds of tiny, calcite-coated lenses.

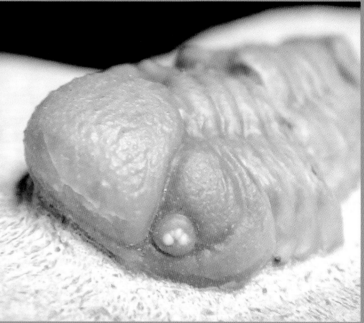

host to have a truly dramatic view of the undersea world around it.

Other Devonian trilobites, such as the Moroccan *Erbenochile issomourensis* and the Bolivian *Fenestraspis amauta* (which possessed the most prominent eyes found in the trilobite kingdom), showcased immense compound eye stacks up to 3 centimeters high. These amazing morphological features were distinguished by over 500 distinct, individual lenses and a pronounced "brim" perched atop each eye column. This characteristic apparently evolved to partially shade the trilobite's impressive assemblage of oculi from the harsh, reflective glare often found in shallow reef environments.

At roughly the same moment in Paleozoic history, trilobite species like *Struveaspis bignoni* had lost virtually all the optical components of their eyes, with their now disproportionately large ocular surfaces being filled by only three to five tiny lenses. This developmental diversity—which stretches back to the Cambrian, when large species including *Conocoryphe sulzeri* appear to have had no detectable eyes at all—seems to reflect the shallow or correspondingly deep-water environments in which these trilobites lived. Those widely varying eye sizes and lens counts may also indicate the role that natural phenomena, such as filtered sunlight and marine turbidity, played in the various ecological niches that these arthropods called home. Unlike the optics of any other organism that has existed on Earth either before or since, the lenses of trilobite eyes were constructed entirely of calcite. This unique evolutionary advancement apparently provided these ancient creatures with an unmatched ability to perceive the light and colors at play in those Paleozoic seas.

When considering the "space-age" symmetry presented by each compound lens, trilobite eyes seem to even further defy their incredibly ancient ancestry, and as we stare into the fossilized eyes of a trilobite today, it takes only minimal imagination to sense that these primeval animals may indeed be staring back at us, providing a dramatic link to life some 500 million years in the past.

RAPID REPORTS

Cortez Mountains, Nevada

In recent years, a high elevation exposure housed within the Lower Devonian Wenban Limestone of Nevada's Cortez Mountains has yielded a modest supply of spectacular trilobite material. These specimens include complete, exceptionally large examples of rare phacopids, lichids, odontopleurids, and dalmanitids. Since the latest explorations of the Wenban's 415-million-year-old layers began in 2009, the recovered trilobite material has been remarkable in both size and quality . . . if not necessarily in quantity. Approximately two dozen complete specimens of the spinose, 12-centimeter-long *Viaphacops clavigar* have been found, along with only a handful of the 18-centimeter *Synphoria nevadensis*, which features a bizarre "inverted" tail. But it is the two distinctly different species of unidentified lichids—one with a bulbous glabella

PHACOPS IOWENSIS (SUBSPECIES)

Middle Devonian, basal Eifelian; Bois Blanc Formation; Southern Ontario, Canada; 6.2 cm

This attractive specimen represents a still undescribed phacopid from a rarely collected formation. The widely spaced lenses on its compound eyes are highly unusual.

and series of lengthy spines ringing its pygidium, the other with pronounced eyes and a huge, half-moon tail—that have attracted the preponderance of interest from both collectors and academics. Some of these lichid specimens range up to 40 centimeters in length, placing them among the largest Devonian trilobites ever unearthed in North America.

Hunsruck, Germany

Since the fourteenth century, thin finely grained sheets of charcoal-gray slate (used primarily as roofing tiles) have been pulled from Devonian-age quarries located throughout the mountainous German region of Hunsruck. Despite world wars and economic downturns, quarrying at the site, commonly referred to in fossil-centric circles as Bundenbach due to that town's nearby proximity, has continued unabated. Although certainly not the primary reason for this mining activity, the appearance of magnificent and academically important fossil fauna has repeatedly drawn international attention to this Rhine Valley locale. Perhaps best known for its diverse assemblage of 400-million-year-old starfish and crinoids, even complete jawless fish have made occasional appearances upon the area's centimeter-thick Hunsruck Slate sheets. The trilobites found here include *Chotecops ferdinandi*, *Rhenops lethaeae*, and *Parahomalonotus planus*, and they are perhaps most notable for their occasional soft-tissue preservation, which has yielded amazing anatomical detail when subjected to modern X-ray technology.

Penn-Dixie, New York

A select number of fossil-filled locations bordering Lake Erie in western New York have long been noted for their exceptional Middle Devonian fauna. Reports regarding 385-million-year-old crinoids, brachiopods, and bivalves date back nearly two centuries. Despite the attention those abundant fossils have long drawn from the local populace, it is the unique accumulation of trilobites that can be found in the region's rich Windom Shale and Moscow Formation limestone deposits that has continually attracted a lion's share of acclaim. Both Eighteen Mile Creek (named for its distance from the Niagara River) and the nearby Penn-Dixie quarry are of singular paleontological interest due to their beautifully preserved examples of such trilobites as *Eldredgeops rana*, *Bellacartwrightia calliteles*, and *Greenops barberi*. The 4-centimeter-long *Eldredgeops* are of particular note because they can occasionally be found in mass-mortality plates featuring dozens of overlapping specimens, a phenomenon some scientists believe may represent mating or molting assemblages. In addition, these locations have proven to be incredibly popular with amateur fossil collectors who visit both Penn-Dixic and Eighteen Mile Creek—often in organized "club" outings—from April until November.

Eifel Region, Germany

Middle Devonian trilobites were initially found in the Eifel Region of Germany's Rhine Valley in 1825. It was then that geologist Heinrich Georg Bronn discovered a small, semienrolled specimen that he named *Calymene schlotheimi*, which has subsequently been reclassified as *Geesops schlotheimi*. Since that time, hundreds of complete and partial examples of that phacopid have been unearthed from this prolific outcrop of the 395-million-year-old Ahrdorf Formation. The fossil-oriented excavations at this site near the town of Gees became so intense during the second half of the twentieth century—including the use of powerful rock-moving machinery—that in 1984 German officials had to intervene and ban all further commercial

diging. By then, however, a full assortment of beautifully preserved, three-dimensional trilobites, representing dozens of species—including such rarities as *Asteropyge punctata, Ceratarges armata*, and *Scutellum geesense*—had already been found. These generally small (2- to 4-centimeter long), brown-to-black toned trilobites continue to be considered the gems of many European museum displays, and they are much-coveted centerpieces for private collections around the world.

LICHID N. SP.

Lower Devonian; Wenban Formation; Cortez Range, Central Nevada, United States; 38 cm

This monster-sized lichid represents one of the largest complete Devonian trilobites ever found in North America.

DEVONIAN TRILOBITE PHOTO GALLERY

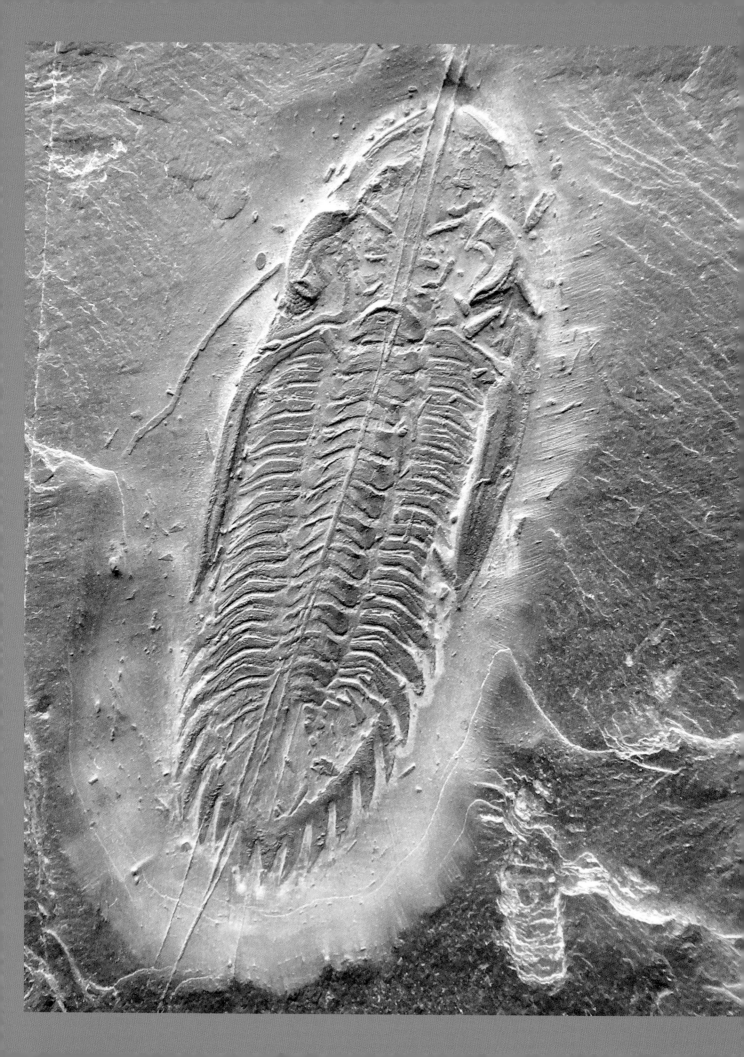

(ABOVE) **DIPLEURA DEKAYI (GREEN, 1832)**

Middle Devonian (Givetian); Skaneateles Formation, Hamilton Group; Delphi Station Member, New York, United States; 20.7 cm

This classic trilobite features impressive size and a thick, black calcite shell. Many known examples of this relatively common species have suffered a degree of torquing due to the planet's internal forces.

(OPPOSITE PAGE, TOP LEFT) **BRAUNOPS SP.1 BASSE, 2003**

Lower Devonian, Lower Emsian; Hunsrück Slate; Bundenbach, Germany; 8 cm

The mudstone layers that are characteristic of this location can occasionally preserve trilobite soft parts, including legs and antennae, although none are visible on this rare specimen.

(OPPOSITE PAGE, TOP LEFT) ODONTOPLEURID N. SP.

Lower Devonian; Wenban Formation; Cortez Range, Central Nevada, United States; 17.3 cm

This large specimen is one of the most impressive trilobites emerging from a recently discovered Nevada locale.

(OPPOSITE PAGE, TOP RIGHT) ODONTOCEPHALUS AGERIA (HALL, 1861)

Middle Devonian; Onondaga Limestone; Needmore Shale Member; Perry County, Pennsylvania, United States; 10 cm

A distinctive cephalic "cowcatcher" (perhaps used to stir seafloor sediments) distinguishes this famous Devonian species.

(OPPOSITE PAGE, BOTTOM) LICHID N. SP.

Lower Devonian; Wenban Formation; Cortez Range, Central Nevada, United States; 15.5 cm

Less than a handful of these large trilobites have been found in a high-elevation outcrop located some 1,500 meters up in the Cortez Mountains.

(ABOVE) MONODECHENELLA MACROCEPHALA (HALL, 1861)

Middle Devonian; Centerfield Limestone Formation, Hamilton Group; Erie County, New York, United States; 5.2 cm

This species was first described by the legendary James Hall. A similar genus, *Pseudodechenella*, also emerges from these same layers.

(LEFT) *PSEUDODECHENELLA ROWI* (GREEN, 1838)

Middle Devonian (Givetian); Centerfield Limestone, base of Ludlowville Formation, Hamilton Group; Erie County, New York, United States; 4.7 cm

With its smooth carapace, this species is easily distinguishable from *Monodecehenella*, a species frequently found in the same formations.

(BELOW) *KETTNERASPIS TUBERCULATA* (CONRAD, 1840)

Lower Devonian; Kalkberg Limestone; Albany County, New York, United States; 4.3 cm

This rare trilobite bears striking similarities to a similarly aged species found in Oklahoma.

(OPPOSITE PAGE) *LICHID N. SP.*

Lower Devonian; Wenban Formation; Cortez Range, Central Nevada, United States; 15.4 cm

Found in a high-elevation exposure, this rare and yet undescribed species exhibits similarities to certain Silurian lichids, but this Devonian example was unearthed in sedimentary layers some 10 million years younger.

(ABOVE, LEFT) *CHEIRURUS SP.*

Lower Devonian; Alnif, Morocco; 8 cm

This is an unusual North African species displaying affinities to *Crotalocephalus*. Studies on Moroccan cheirurids and the nation's Devonian trilobites in general are just beginning to capture the attention of the academic community.

(ABOVE, RIGHT) *BELLACARTWRIGHTIA WHITELEYI* **LIEBERMAN AND KLOC, 1997**

Middle Devonian; Moscow Formation, Windom Shale Member; Hamburg, Erie County, New York, United States; 5.9 cm

This attractive trilobite species has been named in honor of Tom Whiteley, an amateur paleontologist noted for his extensive twenty-first-century research on New York State trilobites.

(OPPOSITE PAGE) *ACANTHOPYGE SP.*

Middle Devonian; Marakib Formation; Jebel Issomour, Morocco; 9.3 cm

A pocket featuring more than 20 of these large lichids was found in 2010, with a high number of complete specimens quickly reaching the commercial market.

(LEFT) *DALMANITES N. SP.*

Lower Devonian; Wenban Formation; Cortez Mountains, Nevada, United States; 14.5 cm

This large trilobite was unearthed in a recently discovered outcrop high up in the Cortez range. Note both the calcite seam that runs through the cephalon and the slightly inverted shape on the pygidium.

(RIGHT) *MYSTROCEPHALA ORNATA* (HALL, 1876)

Middle Devonian; Ludlowville Formation; Livingston, New York, United States; 1.5 cm

Small, ovate, and ornate, this unusual species is one of the true trilo-treasures of the fossil-rich Devonian outcrops of central New York State.

KONEPRUSIA N. SP.

Lower Devonian; Issomour Mountains; Djebel Oufaten, Morocco; 5 cm

The degree of spinosity achieved by some Devonian trilobites was nothing less than astonishing.

CARBONIFEROUS/PERMIAN PERIODS

359–252 Million Years Ago

5

NEW MEXICO AND KAZAKHSTAN: *THE DEMISE OF TRILOBITES*

TRILOBITES EXISTED for nearly 270 million years. And for a significant portion of their sustained stay on our planet, not only did they survive, they thrived. These ancient arthropods filled the world's oceans from the earliest stages of the Cambrian Period, 521 million years ago, until their eventual demise at the end of the Permian Period, 252 million years ago, a time when 90 percent of life on Earth was suddenly and rather mysteriously eradicated. That cataclysmic event, the largest mass die-off in global history, fittingly has become known as the Great Permian Extinction, and it also served as the definitive end line for the entire Paleozoic Era.

Trilobites evolved continually throughout their protracted passage through Deep Time. During that extended journey, they came to occupy a remarkable array of marine habitats, eventually producing more than 180 academically accepted families featuring 5,000 different genera and 25,000 recognized species that displayed a startling diversity of sizes, shapes, and surface ornamentations. By any measurable

AMEURA MAJOR (SHUMARD, 1858)
Late Pennsylvanian (Missouri age); Shawnee Formation; Doniphan Shale bed of Lecompton Limestone Member; Missouri, United States; 3.2 cm

This species ranks among the most renowned American Carboniferous-age trilobites. Collectors from all corners are drawn by its color, shape, and rarity.

criterion, trilobites rank among the most successful organisms ever to appear on our world. Despite their durability and obvious skill for adapting to Earth's ever-changing ecological conditions, however, trilobites underwent a slow yet steady decline before eventually succumbing to a variety of internal and external pressures.

But why did this happen? What caused these highly resilient creatures to dramatically disappear, leaving behind nothing more than their fossilized remains? Could it have been, as some scientists now suggest, something as basic as subtle changes in worldwide sea levels brought on by an early example of global warming? Perhaps it was the result of a series of undersea methane explosions that poisoned oceans around the globe. Or maybe their decline and eventual eradication can be attributed to the rise of fast-swimming predators that viewed trilobites as little more than sushi in a shell.

"Many people think only of the dinosaurs when it comes to a catastrophic natural event terminating a large segment of life on the planet," said Bill Barker, a long-time trilobite collector and dealer living in Arizona. "It seems that similar scenarios have occurred on numerous occasions throughout Earth history, and one of those played an essential role in the end of the trilobites."

It is a relatively straightforward task to trace the rise and fall of the trilobite line through its prolonged Paleozoic voyage. Almost as soon as these creatures emerged in the primal seas, trilobite diversity and distribution was nothing less than astonishing. *Olenellus*- or *Redlichia*-filled biozones can now be found within Lower Cambrian strata on almost every continent, ranging from the Chengjiang deposits of China, to the Kinzers Formation of Pennsylvania, to the Issafen layers of North Africa, to the Emu Bay outcrops of southern Australia. Indeed, the proliferation of trilobite species was never greater than it was only a few million years after their initial appearance.

By the dawning of the Middle Cambrian, some 510 million years ago, trilobites dominated the seas, with thousands of distinct species occupying virtually every available pelagic and benthic niche. At the same time, the multiplicity of predators that feasted on these arthropods was also on the rise. The result was a daily battle for survival that too often did not go the trilobites' way. Some observers have even noted that from this time on it was all downhill for these incredible invertebrates, a long slide that would last for the next 250 million years!

Throughout the Ordovician and Silurian, the number of trilobite genera inhabiting the world's oceans, although still quite impressive, continued to fall. According to the latest scientific evidence, this steady depletion can be attributed to a lethal mix of factors. At varying times, these included everything from a marked decrease in marine oxygen content, to continually shifting global temperatures, to occasional natural disasters—each of which was fully capable of terminating entire trilobite orders. Due to this pernicious combination of environmental elements, by the beginning of the Devonian, 419 million years ago, trilobite species numbered in the hundreds rather than in the thousands. The maritime ecosystems they once commanded were now often subjected to rapidly rising sea levels, which expedited the invasion of jet-propelled cephalopods and terrorizing armored fish, all seemingly determined to cause as much harm as possible on the neighboring trilobite population.

At the dawning of the Carboniferous (comprising both the Mississippian and Pennsylvanian periods in U.S. paleontological parlance), the trilobites'

PILTONIA CARLAKERTISAE BREZINSKI, 2000
Lower Mississippian; Lake Valley Formation; Caballero Mountains, Sierra County, New Mexico, United States; 5 cm

By this late time in trilobite history, only members of the usually diminutive Proetida order survived.

***DITOMOPYGE PRODUCTA* (V. N. WEBER, 1933)**

**Middle Pennsylvanian, Moscovian Regional Stage; Volgograd
Region, Zhirnovsk district, Russia; 3.1 cm**

Tales of complete Russian trilobites from this period were mere
rumors in the West prior to the fall of the Soviet Union. Since
then, dozens of examples have emerged from the nation's
fossil-rich quarries.

decline had become manifest. Their precarious situation had been further exacerbated when a series
of major global extinction events at the end of the
Devonian decimated over 60 percent of life in the
oceans. In the aftermath of these widely destructive episodes, only members of the relatively compact, simplistically designed Proetida order had
endured to complete the trilobites' remaining
100-million-year existence on Earth. Even as their
species count—along with their worldwide population density—dropped precipitously in a post-
Devonian world, trilobites continued to evolve to

meet a wide variety of environmental pressures
and ecological opportunities.

"It's remarkable that trilobites were around for
more than a quarter-billion-years," said Martin
Shugar, a field associate with the American Museum
of Natural History. "It's also apparent that marine
conditions were not always favorable for them. But
for a very long time they adapted and survived."

The trilobite species that lived during the Mississippian, Pennsylvanian, and Permian periods were
certainly not the most morphologically impressive
examples of their class. Yet even considering their
frequently diminutive size (usually less than 5 centimeters) and their modest, ovate body plan—a
design that allowed them to burrow beneath the
seafloor ooze to better avoid both fluctuating
water temperatures and the now constant threat
of predation—these trilobites still contributed
significantly to their kind's lingering legacy. They
may have been far removed from the sleek, often
elegant species that dominated the Cambrian, or
the 30-centimeter spinose giants that inhabited
the Ordovician, Silurian, and Devonian seas, but
in looks, design, and lifestyle they were still very
much quintessential trilobites.

Despite being in such steep decline, the fossilized evidence of these last in line trilobites can
be found in numerous localities across the face of
the planet. The 350-million-year-old Mississippian-age outposts of Missouri, for example, have
long been a favorite of collectors in search of such
small (1- to 2-centimeter) species as *Ameropilto-
nia lauradanae* and *Comptonaspis swallowi*. These
captivating specimens, drawn from the fossil-rich
Chouteau Formation, are noted for their finely
detailed, three-dimensional preservation, a feature
that has made some dedicated midwestern enthusiasts focus almost exclusively on their acquisition.

Over the past two decades, New Mexico has also
emerged as a Mississippian Period epicenter. Twenty-
three trilobite species have been identified from the

neighboring Caballero and Lake Valley formations, including *Piltonia carlakertisae, Namuropyge newmexicoensis*, and *Pudoproetus fernglenensis*. These trilobites, which can reach up to 6 centimeters in length, are often preserved in a dark brown or black calcite that contrasts vividly against a reddish-pink matrix. This attractive coloration, as well as their increasing scarcity on the commercial market, has made these southwestern proetids popular additions to many private collections. Another important Mississippian-age trilobite site is located in the outskirts of Antoing, Belgium, a picturesque town dominated by towering church steeples and twelfth-century castles. Well-preserved, lightly mineralized examples of such 2-centimeter-long species as *Piltonia kuehnei, Witrydes rosmerta*, and *Bollandia globiceps* have been discovered amid the area's black mudstone outcrops for more than 120 years.

At the start of the Pennsylvanian Period, roughly 323 million years ago, trilobite numbers continued to ebb in the fossil record, but the Proetida order managed to produce an interesting array of species, including such notable North American varieties as *Ditomopyge olsoni, Ameura major*, and the pustule-covered *Brachymetopus nodosus*. In recent years, a small number of 320-million-year-old trilobite genera

DITOMOPYGE SP. (V. N. WEBER, 1933)

Uppermost Pennsylvanian, Upper Carboniferous (Stephanian); Dzhezkazgan Region; Ulutau Mountains, Karagandy Province, Kazakhstan; 2.7 cm

These trilobites, drawn from fertile outcrops in the distant Ulutau Mountains, are almost always found as internal molds and are missing their outer shell.

from the remote Ulutau Mountains of Kazakhstan have invaded the world market. Many of these dolomitic specimens (including 3- to 4-centimeter examples of *Ditomopyge kumpani* and *Griffithides praepermicus*) have been fossilized alongside other fauna, including brachiopods and crinoids, providing a compelling view of life in the seas at this late stage of the trilobites' evolutionary game.

By the time the Permian Period began some 298 million years ago, the limited available fossil material provides stark confirmation that trilobites were barely hanging on in their constantly changing marine world. Their average size had shrunk to only a few centimeters, and their speciation had reached a critical low. Despite all such apparent tribulations, the fossilized distribution of species, including *Paraphilipsia sp.* and *Acropyge multisegmenta*, is still rather impressive, with the often disarticulated remains of these terminal trilobites being found in sedimentary layers in such currently far-flung locations as present-day Hungary, Timor, China, Pakistan, Russia, and Japan.

"Post-Devonian trilobites are usually not the most exciting for collectors," said Barker. "Whether they're Carboniferous or Permian, they all tend to be similar in their appearance. But they should not be overlooked because they do tell an important story, especially about the decline and end of the trilobites."

Whatever elemental factors may have served as the root cause behind their measured but unmistakable downturn—natural selection, climate change, celestial gamma ray bursts, undersea methane eruptions, or even problems caused by their increasingly outdated molting behavior—the fossil record clearly shows that the entire trilobite class was in dire evolutionary straits well before its eventual demise. However, ascertaining exactly which environmental forces served to finally push these incredibly adaptable arthropods beyond the brink of recovery remains something of a Paleozoic mystery. One well-considered recent theory postulates that a string of supersized volcanoes simultaneously erupted 252 million years ago throughout the region that is now Siberia. This unprecedented degree of seismic activity—which may have persisted for years, if not decades—filled both air and sea with enough carbon dioxide residue to slowly suffocate terrestrial and marine lifeforms everywhere on Earth.

When it comes to the termination of the trilobite line at the end of the Permian, however, all our previous conjectures pale in the shadow of the lumbering *Anomalocaris* in the room—the strong possibility that something as drastic as a meteorite strike may have served as the proverbial final straw that broke the trilobites' calcified exoskeleton. Indeed, some academics now believe that an asteroid or meteorite shower—possibly consisting of up to a dozen large-scale, iron-nickel-alloy space rocks that together would have unleashed the power of over a million nuclear warheads—may have simultaneously struck Earth at this critical juncture in the planet's history. If so, the results were uniformly and understandably devastating for life around the globe.

Even in light of steadily growing stacks of suggestive evidence, science has yet to uncover the definitive "smoking gun" that conclusively signals the close of the Paleozoic Era—the kind of irrefutable proof provided by the meteorite-generated crater off the Yucatan Peninsula (as well as the subsequent worldwide iridium layer) that is indicative

PUDOPROETUS CF. FERNGLENENSIS (WELLER, 1909)

Lower Mississippian, Upper Osagian; Edwardsville Formation; Montgomery County, Indiana, United States; 5.2 cm

Found in a variety of outcrops across the United States, this genus exhibits many of the classic features of the Proedita order. This specimen is slightly enrolled, with its pygidium tucked under the thorax.

of the global damage that 65 million years ago dealt the dinosaurs their final blow. It was that calamitous event that suddenly, dramatically, and emphatically marked the end of the Mesozoic Era.

Despite this apparent lack of corroborative material, paleontologists now know that a worldwide disaster of epic proportions also rocked the Earth some quarter-billion years ago, in the process causing the largest mass extinction in the planet's 4.54-billion-year history. More than 96 percent of all marine species and 70 percent of terrestrial life-forms perished in that event's prolonged wake. For a trilobite class already in steep decline, such a devastating development would have proven nothing less than a total catastrophe. Although it seems highly unlikely that these intriguing invertebrates could have survived such a cataclysmic natural disaster even at the peak of their evolutionary powers, their diminished numbers and already antiquated lifestyle made them immediate candidates to join the planet's ever-expanding list of "failed experiments."

As life on Earth has continually shown, the downfall of one species often provides an ideal opportunity for the ascendance of another. Make no mistake about it, much as the eventual rise of mammals owes a great deal to the sudden demise of the dinosaur, dinosaurs themselves owe a great deal to the Great Permian Extinction, an event that effectively cued the end of the Paleozoic and the beginning of the Mesozoic—the Age of Dinosaurs.

It is not merely wild speculation to assume that life on our planet would look radically different today if the numerous mass extinctions that dot the last half-billion-year span of our shared history had not occurred. Who knows, with a few well-placed twists of fossil fate, perhaps some particularly clever trilobite offshoot could have figured out how to traverse the tricky paths of evolutionary change and would now stand (or more likely crawl) as the dominant form of life on Earth.

TRILOBITE MULTIPLICITY

As you peruse the photos adorning these pages, the temptation may well be to envision trilobites as solitary organisms, proud denizens of the world's Deep Time seas captured for eternity in fossilized form as they proceeded stoically, and individually, about their daily routine. After all, the overwhelming majority of those pictured are solo specimens, many rather aesthetically posed amid a matrix field comprised of half-billion-year-old primordial ooze turned to stone.

The truth is, however, that certain aspects of trilobite behavior may not have been exactly as we've long imagined. Indeed, myriad lifestyle characteristics employed by these intriguing invertebrates may have been quite contradictory to our modern-day perceptions. As some within the academic world have recently begun to learn, rather than being isolated loners, many—if not most—trilobite species were highly communal animals. They often congregated in tightly packed groups, perhaps even traversing the world's seafloors in long, single-file, cephalon-to-pygidia "conga" lines, as one 2016 scientific paper colorfully postulates. This was apparently a behavior designed to provide safety in numbers, subsequently furnishing a marked increase in each trilobite's procreative possibilities.

In some notable cases, particularly trilobite fossils found as part of sedimentary outcrops in such now diverse locations as the Czech Republic, Russia, Portugal, Morocco, Utah, and British Columbia, these primal creatures appear to be pervasive

ONNIA SP.
Upper Ordovician, Ashgill Series; Upper Ktaoua Formation; Blekos, Morocco; average size: 3 cm

Plates such as this provide a degree of insight into the possible lifestyles and habits of these ancient arthropods.

within their ecosystem. In these locales (as well as in an ever-increasing number of sites worldwide), layers of Paleozoic rock have been unearthed that are virtually covered in fossilized trilobite remains. Some scientists currently believe that these mass mortality assemblages—such as those dramatically exhibited by the Ordovician asaphid *Homotelus bromidensis*, where hundreds of complete, 6-centimeter-long trilobites have been uncovered fossilized side-by-side, and one atop the other within Oklahoma's Bromide Formation—may reflect the end result of an ancient tidal basin draining or evaporating, leaving its inhabitants quite literally high and dry.

Other experts state that some trilobites, such as the phacopid *Eldredgeops rana*—discovered within the rich Devonian outcrops of upstate New York—may have followed a life cycle that would have drawn their species together in prolific numbers at certain times of the year, creating mating or molting conglomerations. The net effect of such behavioral activities may well have left Paleozoic seafloors carpeted with nearly complete trilobite carapaces.

One interesting theory that has gained traction in recent years states that trilobite species found in the same geological horizon may have lived in numbers somewhat akin to the predator/prey ratios we see today in the wild. Anyone who has ever viewed a *Nat Geo* special on the African savannas knows that prey animals like zebra and wildebeest far outnumber the lions and cheetahs that pursue them. Some 500 million years ago, during the Cambrian, a similar situation may have

existed between the possibly predatory trilobite *Olenoides nevadensis* (an uncommon find) and the prolific *Elrathia kingii* (numbering in the tens of thousands) that they may have feasted upon.

Although trilobites represent relatively rare remnants of an incredibly distant time, as we continue to explore the various sedimentary strata that comprise the Earth's surface, we find that their fossilized exoskeletons are quite ubiquitous. Indeed, through the occasionally pervasive manner of their preservation, trilobites are beginning to reveal tantalizing bits of information concerning their lifestyle in those ancient seas, hundreds of millions of years in the past.

TRILOBITE DESCENDANTS

Every week, rain or shine, a post called "Trilobite Tuesday" appears on the official Facebook page of the American Museum of Natural History. Along with a stunning photo of a rare trilobite species drawn from some distant corner of the globe, there is usually a pithy blurb—written by yours truly—explaining exactly why such a specimen is of paramount paleontological importance. Despite the decidedly out of the mainstream status that trilobites enjoy, the museum's weekly Facebook presentation manages to generate a more than respectable reader reaction. Hundreds of "likes," a goodly number of "loves," even the occasional "wow," along with a score of generally favorable comments routinely dot the space directly below each of these entries. Indeed, such a response has continually marked "Trilobite Tuesday" as one of the AMNH's most consistently read—and commented upon—Facebook offerings.

Amid the expected assortment of "they're cute" or "they're hideous" pronouncements—and accompanying emojis that those on social media seem so determined to share with the world— these trilobite-centric posts on occasion generate

LONCHODOMAS (AMPYX) SP.
Lower Ordovician; Ouled Siimane; Zagora, Morocco; largest trilobite: 2.7 cm

The cephalon to pygidium alignment assumed by these trilobites may reflect the manner in which they traversed the Paleozoic seas.

***GABRICERAURUS DENTATUS* (RAYMOND AND BARTON, 1913)**

Upper Ordovician; Bobcaygeon Formation; Deseronto, Ontario, Canada; 10.4 cm

Although trilobites left no direct descendants, many who view specimens such as this note striking similarities to some contemporary arthropods.

comments that merit both consideration and reply. Quite possibly, the most repetitive among these is the surprisingly hard to dismiss assumption that various contemporary creatures—whether pill bugs, isopods, or horseshoe crabs—must in some way, shape, or calcite-coated form be direct descendants of the noble trilobite line. Perhaps this is the appropriate space to put such an erroneous notion in its proper Paleozoic place. Quite simply, it is now an irrefutable scientific fact that following their demise at the conclusion of the Permian Period some quarter-billion years ago, the entire trilobite class came to a rather abrupt and eternal end. Despite any lingering conjecture to the contrary, after their incredible journey through more than 270 million years of evolutionary time, these

highly resilient invertebrates left behind no living descendants. None . . . nada . . . zilch . . . zero!

Yes, a variety of contemporary arthropods may share a smattering of morphological features with trilobites, including legs (which were first noted in trilobite fossils in 1870), an easily molted external shell, the placement of their eyes, and even a multijointed thorax. But other equally important anatomical aspects of their exoskeletons, such as eye ridges and the shape of their hypostome, mark trilobites as a totally separate branch of the arthropod family tree. As a handy point of reference, the arthropod phylum is divided into four basic subgroups. The most prominent of these include hexapods (insects such as ants and bees), myriapods (centipedes and millipedes), crustaceans (shrimp and crabs), and chelicerates (spiders, scorpions, and horseshoe crabs). Trilobites are related to all—but are direct ancestors to none.

Why did trilobites leave nary a single descendant in their Paleozoic wake? Other less notable members of the arthropod phylum—including some that initially shared the Cambrian seas with their trilobite cousins—managed to more nimbly navigate evolution's oft-convoluted maze. Even the extinct dinosaur clade left behind thriving evidence of their long ago lives through the presence of modern birds. Why not trilobites? Any answer to this tantalizing yet basic question remains clouded in primordial mystery, but a few theories have been presented in recent years, each designed to shed a ray of light on this timeless tale.

Perhaps the primary reason academics cite for the lack of modern trilobite offshoots is that from the moment of their emergence more than 521 million years ago these were the most ancient members of the arthropod line. The first chelicerates and crustaceans, for example, didn't appear on the marine scene for another 10 to 15 million years. Although trilobites evolved continually throughout their lengthy stay amid the Deep Time seas, their already somewhat antiquated anatomy often made them appear one solid step behind the times when it came to the race for survival. Another rationale for the trilobite line's demise postulates that throughout their Paleozoic passage these unique organisms found themselves playing an increasingly less significant role within the water world that surrounded them. Although thousands of trilobite species dominated the Cambrian seas, by the conclusion of their crawl through time in the Permian, their size, numbers, and dominance had all been reduced to mere blips on the evolutionary scorecard. Their ongoing viability, even prior to the global mass extinction event that dramatically signaled the end of the Paleozoic Era, was, to say the least, tenuous.

When it comes to the untimely termination of the trilobite class, it is apparent that unlike a certain infamous Hollywood character, they *won't* be back . . . and neither will any of their nonexistent arthropod progeny. In the case of this invertebrate inspired Paleozoic production, there will certainly be no sequel.

HESSLERIDES BUFO (MEEK AND WORTHEN, 1870); PLATYCRINITES HEMISPHAERICUS (CRINOIDS)

Lower Mississippian; Edwardsville Formation; Crawfordsville, Indiana, United States; trilobite: 6.2 cm

This creative combination reflects seafloor faunal interaction late in the trilobites' crawl through evolutionary time.

RAPID REPORTS

Crawfordsville, Indiana

The Edwardsville Formation outcrops located near the rural town of Crawfordsville, Indiana, are renowned for their world-famous crinoid assemblages (which feature more than 60 academically recognized species), but these Lower Mississippian quarries have also yielded impressively sized *Hesslerides bufo* trilobites, some up to 6 centimeters in length. On rare occasions, these two faunal elements emerge in close association on the 340-million-year-old limestone slabs, creating a dramatic captured in time look at life in the primal seas. Ironically, those who specifically seek the magnificent, three-dimensionally preserved crinoids, for which this locale has been justifiably lionized since its discovery in 1842, usually show little to no interest in any trilobite detritus that may inadvertently appear on their carefully collected sedimentary stones. A few enthusiasts have even admitted to having the quarry's trilobite matter—particularly if it represents an incomplete specimen—surgically removed from their finds to best preserve the stoic sanctity of their otherwise unsullied crinoid discoveries. Of course, most who stumble upon a collection-worthy *Hesslerides* during their excavations are only too eager to save it, show it off, or sell it to someone who may more fully appreciate its importance as one of the last trilobite species ever to swim through the planet's murky seas.

Lodgepole Formation, Montana

No matter how some among us may choose to recognize, rationalize, and resolve the seismic changes that have transformed the Earth's surface over the last half-billion years, as you meander through the mountainous environs of southern Montana, it's still difficult to fathom that this imposing landscape was once located at the bottom of a shallow, semitropical sea. During the Lower Mississippian, approximately 350 million year ago, the high-altitude surroundings that now comprise rugged Gallatin County were indeed part of a rich, marine shelf environment. That ecosystem proved highly hospitable to a host of proetid trilobite species, including *Brachymetopus* *mccoyi* and *Australosutura sp.*, that represented the survivors of a Devonian-ending mass extinction event that wiped out more than 60 percent of life on Earth, including entire orders of trilobites. Today the fossilized remains of these late-stage trilobites are found at a significant elevation in the tree-lined Centennial Mountain Range. Although merely reaching any of the 2,500-meter-high dig sites can be a yeoman's task—as is extracting these small (generally under 3-centimeter) trilobite remnants from their hard, limestone encasements—the specimens that have been recovered tend to be beautifully preserved in a thick, golden-brown calcite, with many of the key morphological features of their exoskeletons displayed in exquisite detail.

CARBONIFEROUS TRILOBITE
PHOTO GALLERY

(OPPOSITE PAGE, TOP LEFT) ***KASKIA CHESTERENSIS* (WELLER, 1936)**

Mississippian; Big Clifty Formation; Sulpher, Indiana, United States; 2.5 cm

By this late stage in their history, trilobite speciation had reached an all-time low.

(OPPOSITE PAGE, TOP RIGHT) ***DITOMOPYGE ARTINSKIENSIS* (V. N. WEBER, 1933)**

Lower Permian, Artinskian; Arti Formation; Krasnoufimsk; Middle Ural, Russia; 4.4 cm

This intriguing, midmolt example represents one of the larger species of last in line Permian trilobites.

(OPPOSITE PAGE, BOTTOM) ***HESSLERIDES BUFO* (MEEK AND WORTHEN, 1870)**

Lower Mississippian; Edwardsville Formation; Crawfordsville, Indiana, United States; 4.1 cm

The smooth contours of this trilobite's carapace indicate that it was probably at home along the seafloor where it could quickly bury itself in soft sediment.

(ABOVE) ***PUDOPROETUS FERNGLENENSIS* (WELLER, 1909)**

Lower Mississippian; Lake Valley Formation; New Mexico, United States; 6.2 cm

This is a large, impressive specimen of a genus also found in Indiana and Missouri.

(ABOVE) GITARRA GITARRAEFORMIS (GANDL, 1977)

Carboniferous, Visean Stage, Upper Chadian Sub-Stage; Hodder Mudstone Formation; Hodder Valley, Lancashire, England, UK; 3.8 cm

Only in recent years have advanced preparation techniques been able to properly showcase specimens such as this.

(OPPOSITE PAGE, TOP LEFT) DITOMOPYGE SCITULA (MEEK AND WORTHEN, 1865)

Lower Pennsylvanian; Marble Falls Formation; San Saba County, Texas, United States; 2.3 cm

These trilobites can occasionally be collected on the surface as one strolls through the rock-strewn landscape of this central Texas site.

(OPPOSITE PAGE, TOP RIGHT) NAMUROPYGE NEWMEXICOENSIS BREZINSKI, 2000

Lower Mississippian; Lake Valley Formation, Nunn Member; New Mexico, United States; 4.4 cm

Featuring a pustule-covered exoskeleton, this distinctive trilobite hails from one of North America's most prolific Mississippian-age sites. The glabella of this specimen was damaged at some point during the fossilization process.

(OPPOSITE PAGE, BOTTOM) HESSLERIDES ARCENTENSIS (HESSLER, 1962)

Lower Mississippian; Lake Valley Formation; New Mexico, United States; 3.2 cm

A small pocket of these well-preserved proetid trilobites was found in 2008, producing half-a-dozen complete specimens.

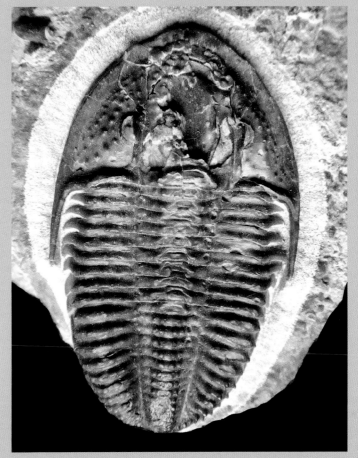

(TOP LEFT) *PALADIN MORROWENSIS* (MATHER, 1915)

Pennsylvanian; Graford Formation; Bridgeport, West County, Texas, United States; 2.8 cm

This is one of the more distinctive Pennsylvanian species. Their brick-red appearance can be attributed to the minerals active during the fossilization process.

(TOP RIGHT) *CUMMINGELLA BELISAMA* HAHN, HAHN, AND BRAUCKMANN, 1985

Lower Mississippian, Tournaisian; Tournai Formation; Tournai, Belgium; 2.2 cm

Unlike most trilobite fossils found in the Tournai location, this specimen possesses a thin calcite shell.

(OPPOSITE PAGE) *KASKIA SP.*

Late Mississippian; Mauch Chunk Formation; Fayette County, Uniontown, Pennsylvania, United States; 2.5 cm

These internal molds represent a rare find in this formation, which has been closed since the 1990s.

TRILOBITE THOUGHTS AND OBSERVATIONS 6

THE TRUTH ABOUT TRILOBITE COLLECTORS

BELIEVE IT OR NOT, there are thousands of people around the globe who collect trilobites with varying degrees of passion. Judging by escalating museum attendance figures and ever-increasing internet search results, many more possess at least a passing curiosity when it comes to the World's Favorite Fossil Invertebrate. One particularly popular trilobite-focused Facebook page recently boasted that it had attained the comparatively lofty level of 10,000 active members—impressive considering the admittedly unusual subject matter involved.

Perhaps the interest of these enthusiasts is limited to possessing a single 510-million-year-old *Elrathia kingii*—the most common trilobite on Earth—a specimen they may have picked up in a museum gift shop during some long-forgotten weekend outing. Maybe their collecting activities have inspired them to gather together representative examples from their home state or region, or from each of the seven geologic periods that comprise the Paleozoic Era. Perhaps they've been resolute enough to construct a medium-sized collection featuring species drawn from each of the 10 recognized trilobite orders—the groupings under which the various families of these ancient arthropods are organized.

Or just maybe they have been motivated by reading *Travels with Trilobites* and have decided to amass key specimens from each of the planet's premier trilobite localities—an effort that could easily catapult their collection total into the hundreds.

CERAURUS PLEUREXANTHEMUS (GREEN, 1832)

Upper Ordovician; Sugar River Formation; Lewis County, New York, United States; 4.5 cm

---

Although a relatively common species, this specimen is one of the most perfectly preserved examples of this legendary northeastern trilobite.

PSILOCEPHALINA LUBRICA HSU, IN HSU AND MA,1948; PSEUDOCALYMENE SZECHUANENSIS (LU, 1962)

Lower Ordovician, Tremadocian; Yinchufu Formation; Liexi, Hunan Province, China; larger trilobite (*Psilocephalina*): 7 cm

In recent decades, an abundance of trilobite material has emerged from behind the previously closed walls of China.

The intent of some collectors is even more grandiose; they may desire to assemble one of the largest, most complete trilobite collections in their city, in their country, or in the world. The objective of these ambitious hobbyists is relatively straightforward. They want to acquire exemplary examples of as many of the 25,000 scientifically recognized trilobite species as is humanly possible. Their goal is made somewhat easier if they're collecting only complete specimens, of which no more than 10,000 varieties are known. However, even the top collectors in the world can rarely brag about possessing more than 3,000 unique species, even if they tend to be "splitters" rather than "lumpers" when it comes to their take on trilobite taxonomy. Whether those trilobites are well-preserved Cambrian ptychopariids from Utah, supersized Ordovician asaphids from Portugal, or concretion-encased Devonian calmonids from Bolivia, these determined collectors want them all.

Many of these tenacious souls will do just about anything in their paleontologically inspired power—be it mentally, physically, or financially—to accomplish this self-appointed task. If they're not scanning fossil websites in pursuit of their next arthropod acquisition, they're flying off to attend rock and mineral shows in various exotic spots around the globe, or at least they are plotting their next trilobite trade with acquaintances who share their paleo-centric predilections. And let's not forget those hearty individuals committed to bringing a far more hands-on approach to their quest for these esteemed remnants of early Earth. Their primary satisfaction derives from venturing out into the field and hunting for their own, self-collected specimens.

"I've discovered that just about everyone I meet has some degree of interest in trilobites," said Sam Stubbs, a lawyer who currently has the core of his 1,500-species collection on display at the Houston Museum of Natural Science. "Once I show people some of the things I've collected, they're always fascinated . . . they just want to learn more. They want to know how I got them, why I got them, where they're from, how old they are, and how much they cost. I guess that's all part of what makes collecting trilobites as enjoyable as it is."

Despite the thousands of casual trilobite enthusiasts who may exist worldwide, when any compendium of *serious* collectors is considered, the number of list-worthy contenders quickly dwindles to a precious few. In sharp contrast to the hobbyist hordes who intently collect everything from classic cars, to baseball cards, to bottle caps, perhaps no more than 100 individuals dotting the face of the planet have the desire, the focus, the contacts, the time, and the economic resources required to assemble a trilobite collection—one featuring anywhere from 300 to 3,000 unique species—that would stand up to the scrutiny and envy of any museum from Boston to Beijing.

There is ample reason for this paucity: the fossiliferous task undertaken by these hobbyists is not particularly easy, either in design or in execution. While some trilobites from perhaps 200 species are readily available to just about anyone who shows an interest in them (check out the 4,000 or more examples of generally common trilo-types for sale on eBay, usually for less than $100), most rare species remain a highly elusive commodity. Access to the best trilobite specimens is often controlled by federal and local restrictions, geographical and geological logistics, as well as by increasingly outrageous prices, which now routinely reach well into four figures for singularly distinctive examples. Despite the roadblocks placed in their paleontological path, however, a significant number of fossil fanatics from all parts of the planet have managed to amass collections featuring thousands of complete trilobite specimens, along with a proportional number of unique species. Rather ironically, some top-tier regional enthusiasts—such as those who specialize in the fabled Devonian phacopids of Sylvania, Ohio—may possess hundreds of outstanding specimens all of one, or perhaps two, trilobite species.

Gathering their material together is only the first step in the collecting process for this small but fiercely dedicated crew of trilobite accumulators. The manner in which they display their holdings is as important and often is as eccentric as the collectors themselves. Some will fill office shelves with their primeval treasures in an attempt to impress friends and business associates. A few will carefully store their trilobites in the basement or garage in hermetically sealed plastic

boxes—sequestered from both harsh sunlight and prying eyes, but sorted, of course, by genus, location, and geological period. Others will place their best examples on prominent display in custom-built, dust-controlled, perfectly spotlighted living room cabinets, ready to share their unusual passion with whomever they deem worthy. A small number of the most accomplished (and fortunate) of these trilobite enthusiasts will align themselves with prominent museums and universities with the objective of seeing their Paleozoic collections eventually form the nucleus of an exhibit that millions will enjoy.

Make no mistake about it, despite what may be the ever-expanding size and scope of their collections, for a preponderance of these elite trilobite "addicts" the hunger to acquire new specimens is never fully satiated. Much like impassioned hobbyists everywhere, no matter how limited their display space or how far their budget has already been stretched, they invariably crave more!

"There is definitely a competitive nature involved with collecting trilobites," said Martin Shugar, whose world-spanning collection features more than 1,000 species. "Each collector wants the best possible specimens, and no matter how many they may own, they always want to attain that next, hard to get one—especially if they know someone else really likes it."

When all is said and done, what exactly is the lure of collecting trilobites? What is it that makes grown men, many of whom are lawyers, doctors, media moguls, or Hollywood heavyweights

DOLICHOHARPES RETICULATUS
(WHITTINGTON, 1949)
Ordovician; Platteville Group; Lancaster, Wisconsin, United States; 4.1 cm

A recent commercial dig at this midwestern site revealed a small number of these attractive trilobites.

***GLYPTAMBON VERRUCOSUS* (HALL, 1864)**
Silurian; Waldron Shale Formation; Shelby County, Indiana, United States; 11 cm

This is a particularly large and impressive example of a rare midwestern dalmanitid. The Waldron layers are world renowned for both their unusual fauna and outstanding preservation.

(Microsoft's Bill Gates and movie land's Nicholas Cage are rumored to have impressive trilobite collections), focus so intently on the Paleozoic permutations of these long-gone marine inhabitants? Some of the ageless attraction of trilobites stems from the recognition that they were among the first dominant forms of multicellular life on Earth, that they survived for 270 million years, and that there was a seemingly endless variety of these primordial arthropods. Trilobites were also perhaps the earliest organisms on our planet to possess complex eyes and calcite exoskeletons, both of which are often beautifully preserved in their fossilized remains. The dizzying degree of morphological diversity that exists within the trilobites' basic three-part body plan also adds immeasurably to their inherent allure; there are spinose lichids, symmetrical asaphids, streamlined harpids, and spectacular cheirurids. And the notion that fossilized trilobite debris has been found in just about every nook and cranny on Earth's craggy surface—from the distant islands of South Australia, to the burning deserts of North Africa, to the rugged steppes of Eastern Siberia, to the majestic mountains of Western Canada—also provides an extra element to their collectability quotient.

"It's a daunting task to even think about trying to collect trilobites from all over the planet," said Carles Coll, a Barcelona-based hobbyist who boasts one of Europe's top collections. "Any time you believe you may be filling in the necessary holes, and picking up the needed pieces, a new site opens in China, Russia, or Bolivia and the work begins all over again."

For many collectors, trying to keep up with the latest in trilobite news, views, and information adds a degree of academically tinged luster to their hobby. Scores of trilobite-related papers, tomes, treatises, and dissertations are released annually by leading institutions and universities. These are gobbled up like manna from heaven—or at least from Harvard—by some hobbyists who pride themselves in constantly staying one step ahead of the competition in their never-ending search for Paleozoic "gold." The work of dedicated amateurs—particularly when it comes to exploring for new fossil-laden localities—has done much to help science better understand the role trilobites played in our planet's evolutionary development. The earnest intent of these enthusiasts to be the first on their block to add the next smooth-shelled Moroccan proetid or big-eyed Russian illaenid to their holdings has helped keep the trilobite-collecting world spinning along at a surprisingly healthy pace from both a financial and a scholarly perspective.

For any number of hobbyists, the realization that so many unusual trilobite species can be obtained via field trips, visits to major fossil shows, and ventures onto the internet can easily turn their collecting aspirations from a mere recreation into what verges on a life's pursuit. Whether a specimen hails from the Cambrian of Sweden, the Silurian of Indiana, or the Permian of Indonesia, there is a collector somewhere who desperately wants to add it to his or her Paleozoic possessions—and a fossil merchant somewhere is just as eager to sell it to them.

Many trilobites exude a certain unconventional elegance with their flowing spines, lustrous calcite carapaces, and geometrically complex ocular symmetry, elements that further enhance their timeless appeal. Most collectors tend to view their trilobites as the somewhat clichéd "butterflies of the sea," but others, not quite as attuned to the perceived wonders of these oft-bizarre invertebrates, are equally inclined to consider them more akin to primordial ocean cockroaches. When it comes to trilobites, beauty is very much in the eye of the beholder.

"Trilobites *are* beautiful," said Shugar. "At least they are to those who appreciate them. I think every slightly overzealous collector has had the experience where a friend comes by and recoils in horror when a prized trilobite is thrust under their nose."

Trilobite fossils and trilobite collectors can both be found on every continent on Earth. North

ZACANTHOIDES GRABAUI (PACK, 1906)

Middle Cambrian, Miaolingian Series, Wuliuan Stage; Langston Formation, Spence Shale Member; Wellsville Mountains, Box Elder and Cache counties, Utah, United States; 6.3 cm

Sleek, streamlined, and spined, this distinctive species is one of the true Paleozoic prizes found within the Spence Shale exposures that dot the rugged Wellsville Mountains.

American enthusiasts appear from coast to coast, with many—especially in the Midwest—specializing in local fare, such as the exceptional examples derived from the famed Ordovician outcrops of Kentucky, Ontario, and Wisconsin. Major European collectors maintain a distinct worldview when it comes to their arthropod acquisitions, though some seem to possess a special affinity for fossiliferous material first discovered by Continental paleontological pioneers a century or more ago. Collectors in Japan have become renowned for their finicky tastes where the inherent artistry displayed by a key specimen may be more valued than its rarity. They are also recognized for their willingness to spend what might initially appear to be exorbitant sums to acquire their target trilobite. And to take it all to the next somewhat illogical level, private museums dedicated to trilobites (and created by the collectors themselves) have recently sprung up in such diverse locales as Mexico, Spain, and Morocco.

For those drawn to this fossil-filled universe, there is an inherent fascination in holding in one's hands the remnants of the world's first significant hard-shelled organism—animals that filled the oceans hundreds of millions of years before dinosaurs set foot on Earth. There is also an inexplicable thrill associated with venturing into the field and uncovering a creature that no human eye has seen before—a specimen from a previously unknown trilobite species or one that might prove invaluable to science's evolving understanding of life's initial, tentative steps on our planet.

Whether you dig deep into sedimentary soils in pursuit of trilobite treasures, barter for these ancient relics with fossil-loving comrades, or purchase them with a mouse-click on the internet, one fact seems abundantly clear—trilobites rank among the most captivating, diverse, and successful life-forms ever to exist on our planet. In certain ways, those primal creatures share at least some of their quirky qualities with leading trilobite enthusiasts around the globe, many of whom expend an extraordinary amount of time, effort, and financial capital in their seemingly tireless quest to collect these amazing arthropods. In the hearts and minds of these dedicated hobbyists, perhaps no time in the Earth's long history was more significant than 500 million years ago—the time when trilobites ruled the seas.

"Trilobites are simply amazing," Stubbs said. "They're beautiful, and endlessly intriguing. They never cease to astound and entertain me. It's hard to explain why . . . they just do. I guess that's what makes me a collector."

FINDING A TRILOBITE

It's convenient to imagine that all we need do to find a collection-worthy trilobite is stroll by an appropriately aged sedimentary outcrop. There, a quick perusal of the surrounding matrix and perhaps a well-placed hammer blow will result in a successful conclusion to our fossil-inspired expedition. Ahhhh . . . if only it was that easy. Discovering a trilobite usually takes quite a bit more effort than some may realize, especially those of us who collect primarily via trolling the internet or visiting local rock and mineral shows. Uncovering a trilobite fossil amid the wonders of nature is an entirely different kettle of arthropods. In most cases, field finds occur in one of three distinctly different ways: you happen to accidentally come across a fossil-bearing locale; you research, explore, and uncover a prime Paleozoic outcrop;

PSEUDODECHENELLA ROWI (GREEN, 1838)
Middle Devonian; Hamilton Group; Warsaw, New York, United States; largest trilobite: 5.1 cm

Uncovering a dramatic triple such as this rates among the ultimate trilobite thrills for a collector.

PARANEPHROLENELLUS KLONDIKENSIS WEBSTER, 2007

Lower Cambrian, Dyeran; Pioche Formation, Delamar Member; Klondike Gap, Lincoln County, Nevada, United States; Each trilobite: 3 cm

In the field, multiple specimens are often covered by matrix. It is only after careful preparation that the true scope of such a Paleozoic treasure is revealed.

or you discover trilobites as part of a residual mining or quarrying endeavor.

Perhaps the classic example of the first scenario occurred when Charles Walcott stumbled on the hallowed Burgess Shale in British Columbia in the early years of the twentieth century. Legend has it that while traversing a narrow path through the mountains during his search for the fossil site (first recognized in 1886 during the construction of the Trans-Canada Railway) the celebrated paleontologist dismounted his horse to examine the animal's injured foot. In doing so, he spied a trilobite-bearing chunk of matrix—quite possibly a partial *Olenoides serratus*—that had conveniently tumbled down the cliff before coming to rest in perfect viewing position. Of course, nobody will ever know just how long that particular piece of Paleozoic sediment may have been lying on the side of Mt. Stephen. But its discovery motivated Walcott to begin extensively exploring the adjoining mountainside until he eventually found the thin band of Middle Cambrian shale from which the original trilobite had emerged. The rest, as they so often say, is history.

The second most likely manner of uncovering a trilobite in the field blends a solid degree of scientific research and a modicum of luck with a required dose of backbreaking labor. In these cases, a scientist or amateur collector may begin the quest by examining old geology books, field reports, or topographic maps—with the intent of uncovering a likely fossil-filled sedimentary outcrop. This knowledge then needs to be put into direct action in the field. An intrepid adventurer may spend countless days, weeks, or even years traversing rugged, off the beaten track terrain looking for that promising layer of fossiliferous rock. And once found, the *real* work begins.

The third of our fossil-finding scenarios involves the exploration of an established mine or quarry, where the discovery of trilobites is usually nothing more than an unexpected—and often unwanted—by-product of that facility's primary endeavor. Such locations as the Valongo outcrops of northern Portugal, the Dudley Formation of the English Midlands, or the Silica Shale of Sylvania, Ohio, have long served their local communities by yielding siltstone slabs or limestone blocks to be used as roofing tiles, construction fodder, or roadway filler. Often embedded within these rich layers are complete fossilized ecosystems and biozones occasionally brimming with trilobites. Understandably, those who first encountered trilobites in these sites back in the eighteenth and nineteenth centuries hadn't a clue as to what they were observing. Indeed, they frequently referred to these strange, calcite-coated shapes in the rocks as "frozen locusts" or viewed them as mysterious fetishes that had somehow become hidden within the surrounding stone.

Thus, there are any number of effective ways to discover a trilobite in the wild, and many of these often hard to reach locations provide even slightly out of shape weekend warriors with a solid chance of landing a true Paleozoic prize. Although visiting these remote repositories may not be easy, any resulting fossil specimen is guaranteed to be a fascinating and treasured consequence of such an arduous outdoor endeavor.

FOSSIL SHOWS AROUND THE WORLD

They come from near and far—from Tokyo, London, and New York, from backwater burbs in Oklahoma, and rural communities in Sweden—all drawn by the lure of rare minerals, out of this world meteorites, and the most impressive display of trilobites on Earth. They come by plane, train, and automobile (with a smattering of doublewide motor homes and bad-boy motorcycles thrown in for good measure) to partake in a veritable feast for the eyes and a potential famine for the pocketbook. They are merchants, scientists, museum officials, casual observers, and dedicated hobbyists, and all have one thing in mind—to either buy or sell the best available specimens for the best possible price.

Over the past three decades, fossil and mineral shows have grown exponentially in size. Many now rank among the most important cultural and financial events on the calendars of their host cities and are among the more intriguing destinations for those wanting to experience the latest and greatest that the geological universe has to offer. Despite the unpredictability of the worldwide economy and the expanding role of the internet, which has cut into the base appeal of these natural history extravaganzas by making exceptional material available to anyone anywhere, these shows remain a prime annual stop off for academics, collectors, and those who find, prepare, and sell everything from trilobites to tektites. There, enthusiasts can enjoy a unique sense of camaraderie, joining together in one place at one time to share their mutual "rock star" fascination.

Some venture to these events to proudly display their latest discoveries—the results of a callous-causing season in the field or a budget-busting outing to some exotic locale. Others show up to frown upon this "gross commercialization" of the natural sciences, believing that the ever-escalating costs associated with the acquisition of prime fossils, minerals, and meteorites have priced such pieces outside the realm of the academic community. Still others gather simply to gawk and stare, overwhelmed by the dizzying diversity of rare material that arrives on a daily basis from every corner of the globe. There is plenty of room at these events for all such widely varying paleontological perspectives to flourish.

These days fossil shows are omnipresent—held in tiny towns on picturesque French mountainsides, in highway-hugging motels in the American Southwest, and in the heart of bustling metropolises in Europe and Asia. Whether you are young or old, rich or poor, a Paris slick or an Ozark hick, you're welcome to attend . . . and to spend. Every fossil and mineral show—whether it takes place in Tucson, Tokyo, Munich, or little St. Marie aux Mines—is the natural history world's equivalent of the United Nations . . . or at least the Tower of Babel. A stroll down any dimly lit motel aisle or tent-lined outdoor causeway is liable to bring you face-to-face with merchants from China, Germany,

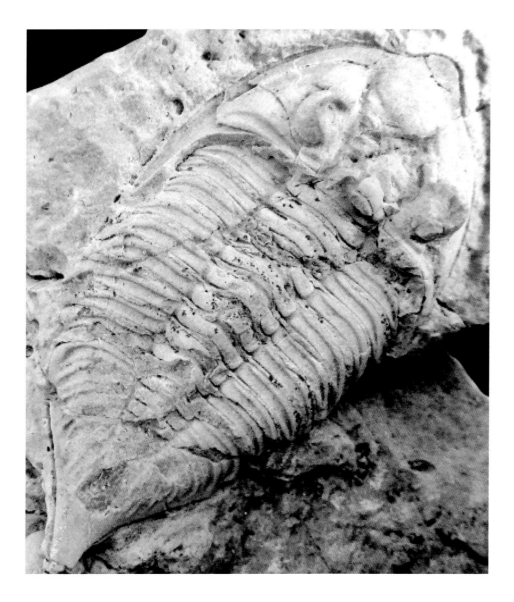

Morocco, Australia, Russia, Canada, and Bolivia, each of whom seems determined to reflect their nation's ancestral ability to either hard-sell or soft-sell the variety of goods they have available. Some of these merchants expect, even want, to haggle over price (in any number of languages); others display outright disdain toward even considering such an unsavory practice. Equal parts carnival sideshow, fossil swap meet, and scientific convention, these natural history events annually draw tens of thousands to their clarion call, often making them the most prominent, exciting, and lucrative tourist attraction their surrounding communities experience all year.

"The action in Tucson always starts in late January, and it is not only an incredible event for collectors, it's also the biggest tourist draw the city has," said Bill Barker, a Tucson resident and one of the top fossil

dealers at the show. "That's saying something when you realize that Tucson is not only a hub for spring training baseball but also hosts a variety of rodeo events that bring in huge crowds."

It is stated by the Tucson tourist bureau that each year over 100,000 visitors attend that city's fossil and mineral show, making it the largest event of its kind in the world. Of course, the fact that the Tucson extravaganza traditionally plays out over a three-week period, whereas most other such shows—which usually take place in summer or early fall—are generally limited to three- to seven-day runs, helps this Arizona exposition stand as the planet's most impressive natural history gathering in terms of both attendance and profitability.

Rumors persist that as much as $30 million can change hands during a typical Tucson show. Many of the high-ticket transactions—which

RAYMONDITES SPINIGER (HALL, 1847)
Middle Ordovician; Decorah Shale; Lincoln County, Missouri, United States; 5 cm

This genus is more commonly found in Ontario, Canada. Complete specimens from this midwestern location are quite unusual.

can range from the sale of complete *T. rex* skeletons to weighty gold nuggets—happen late at night, behind closed doors, in small, crowded showrooms lined with carefully displayed rows of exotic trilobites, opalized dinosaur bones, world-class minerals, and glittering ammonites. These quickly constructed places of business are nothing more than typical motel rooms where the beds have been removed (or converted into somewhat wobbly display areas), and track lighting has been jerry-rigged to create a more appealing ambiance in which to view prospective purchases. In direct accord with the town's unwritten and unspoken "anything goes" show attitude, these hotel rooms, which during the other 49 weeks of the year struggle to sell for $79 a night, go for a cool $229 throughout the run of the Tucson gathering.

Unlike similar shows in Shanghai, Stuttgart, and Denver, where such annual events are generally absorbed by either the size of their host city or the less-centralized location of the show setting, the fossil and mineral spectacular in Tucson has a major impact on the community's economy. Entire areas of downtown are dedicated to show displays, and roadside advertising, which seems to start as soon as you leave the airport, directs one and all to the dozens of free-admission satellite shows held in and around town. Indeed, it is this impressive jumble of generally unsanctioned events (with the only "true" Tucson Fossil and Mineral Show being held at the city's Convention Center over a three-day period in mid-February) that many attendees consider to be the *real* Tucson exposition.

Featuring hundreds of dealers setting up shop in rooms often so cluttered with material that it spills out onto the front porch or adjacent parking lot, a stroll through any of these motel shows offers an exciting—and highly unpredictable—opportunity to wade through some of the greatest treasures in the natural history world.

In sharp contrast to most museum displays, where items such as rare trilobites, meteorites, and crinoids are usually sequestered behind thick

METOPOLICHAS PLATYRHINUS (SCHMIDT, 1907)

Middle Ordovician, Llanvirnian; Kunda Regional Stage (Hunderum-Valaste); Sillarou-Obuchov Formations; Volchov River Valley, St. Petersburg, Russia; 10.1 cm

One of the strangest—and rarest—lichids emerging from the fossil-rich rocks of western Russia, academics are still debating the possible role this species' distinctive "snout" may have played.

TAIHUNGSHANIA BREVICA (SUN, 1931)

Lower Ordovician; Meitan Formation; Zunyi (Tsunayi), Guizhou Province, China; 8.6 cm

This unusual species was acquired from a Chinese merchant at the Tucson Show in the late 1990s.

panes of protective glass, in Tucson visitors can get up close and personal with the objects of their Deep Time desire. You can pick up and examine just about any available piece; but be careful—if you break it, you take it!

Successfully navigating through the intricate maze of delicate mineral crystals and fragile fossil fish that line many of these motel rooms is a challenge unto itself. One second you may be gingerly stepping over an assemblage of Jurassic-age Chinese reptiles left haphazardly on the floor; the next moment you're circumventing a chest-high display of 100-million-year-old German mosasaur bones that the dealer has placed in the hallway. To the uninitiated, these shows often appear to be one small step removed from organized chaos. Beer flats full of rare aquamarine crystals from Afghanistan rest gingerly on broken-down tables, and interested observers jostle with one another in a room's narrow confines to be the first to grab hold of a particularly precious trilobite from Oklahoma. For most show-goers, such a bizarre *and* bazaar atmosphere is all just part of the fun, although many attendees would grudgingly admit that a very fine line separates love and hate when it comes to dealing with the out-of-control aura that can, at times, surround the Tucson experience.

"Going to Tucson each year is a rite of passage for any serious collector," said Tom Lindgren, who has been a top fossil dealer at the show for more than three decades. "It's a crash course in not only learning what's available but also how to find it . . . and how to buy it."

For all the fun, frenzy, and fossils that highlight a visit to the Tucson show, a darker element has begun to infringe on what transpires not only in Arizona, but at each of these planet-spanning events. In recent years, as more fossil material with potential scientific significance has emerged upon the world stage, increased attention has been placed on the legality, and potential lack thereof, that surrounds these pieces. With an escalating number of nations—including Canada, Portugal, Argentina, China, Germany, Mongolia, Australia, the Czech Republic, and the United States—introducing "cultural heritage" and "national artifact" legislation designed to restrict the collection, export, and sale of its natural resources (including some fossils, gems, and minerals), a more intense light is being focused on many of the specimens being offered for sale at these shows. This spotlight has also begun to cast its revealing glow on the numerous natural history auctions now held annually around the globe by the likes of Sotheby's and Bonhams—firms that intently scour every Tucson, Munich, and Tokyo showroom for material worthy of being featured in their next fossil-filled bidfest.

Whether shouted or whispered, more and more questions are being asked by institutional officials, knowledgeable collectors, and government representatives. Was that 66-million-year-old *Triceratops* skull from Colorado found on public or private land? Does that dinosaur egg nest from China have proper documentation? Shouldn't that Burgess Shale *Sidneyia inexpectans* be in the Smithsonian? Should that strange-looking mammal from the Messel deposits of Germany be available for sale at the jaw-dropping price of $1 million? These queries, and many more, have begun to dominate conversations at fossil shows . . . with good reason.

TRICOPELTA BREVICEPS (HALL, 1866); *SALTERASTER GRANDIS* (STARFISH) (MEEK, 1872)

Upper Ordovician, Katian; Cincinnatian Series, Richmondian Regional Stage; Liberty Formation; Preble County, Ohio, United States; trilobite: 2 cm

This is a rare trilobite species seemingly interacting with a large starfish. When fossils like this are found, they provide a unique window upon a long ago world.

In recent years, seizures have been made in Tucson for the illegal importation of Devonian fish specimens from Australia, and numerous shipments of mineral and vertebrate fossil material from China and Mongolia have been impounded by U.S. authorities at the behest of each nation's respective government. Many American merchants have been forced to become aware of legislation like the recent Omnibus Land Preservations Act, which makes it illegal to sell—or buy—even some common fossils found on federal land. The results of such actions and reactions have made at least some show-goers acutely aware of what they're purchasing, and from whom they're purchasing it. That is especially true when one considers that consumer "ignorance" is apparently no longer a viable excuse when it comes to breaking any of these newly fashioned laws—many of which have supposedly been designed to protect the fossil market from itself.

"It's now up to the person who's buying to be as knowledgeable as the dealer when it comes to what can and can't be sold," Barker said. "Often at shows like those in Tucson, you'll walk around and see specimens that you *know* shouldn't be there for one reason or another."

Aside from the ever-increasing number of legal questions that now surround fossil and mineral shows around the world, another concern has been growing among many who routinely attend these events: that the power of the internet will continue to usurp or at least diminish the role such gatherings serve in the promotion and sale of natural history items. After all, why undertake the time, effort, and expense of traveling to a show perhaps half a world away when you can sit comfortably at home and view similar specimens available for purchase on any number of fossil-oriented websites? Could it be that unique events such as the Tucson Fossil and Mineral Show might soon follow the trilobites' path to extinction? For those

whose livelihoods depend on marketing natural history items, this has become a concern that most involved with these high-profile, paleontologically tinged expositions never thought they'd have to seriously consider.

"Even just a few years ago, the idea of the Tucson show not going on every year would have been absurd to me," said Barker, who has been participating in the event for the past 40 years. "But a number of factors have all begun to impact its ongoing viability. But I'd hate to see it go. And in all honesty, I feel confident that the Tucson show will be around for a long time to come."

TRILOBITE RESTRICTIONS: *RULES AND REGULATIONS*

Over the past decade, a surprising and somewhat alarming trend has started to spread across the globe. During that time, an increasing number of government agencies have begun enacting legislation expressly designed to place fossils—including, in some cases, trilobites—on wide-ranging "restricted" lists. Such judicial actions have gone to great lengths to mark these fossils as "national treasures" or "cultural artifacts," in the process banning—or severely limiting—their collection, sale, or trade. In the course of implementing their protective practices, many of these bureaucratic bodies have even tried squeezing fossils into already existing laws designed to safeguard their homeland's cherished archaeological discoveries, with the net result often being an uncomfortable display of square peg in a round hole politicizing. Everywhere from Argentina and Brazil, to Germany and Scotland, to Australia and North Africa, legislative branches are tightening their grip on their nation's prized paleontological reserves, making such items increasingly difficult to buy and sell, or even to study.

Exporting fossils is now basically banned throughout China, and stories surrounding the

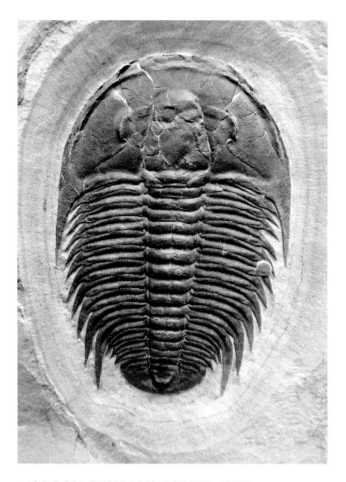

MODOCIA TYPICALIS (RESSER, 1938)
Middle Cambrian, Series 3, Drumian; Lower Marjum Formation;
House Range, Utah, United States; 5 cm

If certain recently written laws are fully enforced, even classic trilobites like this one may soon become rarities.

illegal sales of Mongolian *Tarbosaurus* dinosaurs—involving arrests, convictions, and reparations—have recently filled the international press. Morocco and Russia, which over the past three decades have emerged as the dual lynchpins of the commercial fossil trade, have also begun to toughen their domestic paleontological policies. The combined impact of these efforts has served to significantly limit the supply of top-grade fossil material, including trilobites, entering the world market. At the same time, freshly instituted laws have begun to severely curtail the access everyone from local merchants to museum-affiliated

academics had gained to specimens found within the borders of nations ranging from Mexico to the Czech Republic. Even in the United States and Canada, countries with rich paleontological traditions ranging from pioneering work with dinosaurs to landmark discoveries involving primitive, soft-bodied arthropods, an abundance of fossil-oriented twenty-first-century legislation has been passed. Whether through design or happenstance, these laws have served to significantly limit the ability of nonauthorized personnel to search for, dig, and distribute fossil material found on federally controlled land. While certain restrictions have long been in place for vertebrate fossils, especially dinosaurs, the inclusion of invertebrate specimens such as trilobites within these statutes adds a confusing and somewhat unexpected twist to this still-evolving scenario.

As one might imagine, the fossil-inspired regulations now sweeping the planet have ignited a firestorm of debate between those who collect and those who wish to curtail such collecting practices. Many hobbyists—whether they reside in Pioche, Nevada, or Paris, France—have never seriously had to consider the legal ramifications surrounding their acquisition of a fossil such as a trilobite. Yet collectors now occasionally find themselves asking—or being asked—whether that rare Marble Mountains *Bristolia insolens* sitting on their living room shelf rightfully belongs to them, or in the nurturing grasp of a museum, a university, or a government agency. It is a controversy that often stirs the souls of those in the collecting community. Many enthusiasts believe not only that the possession of a potentially valuable and scientifically important trilobite is their prerogative but that it is also their responsibility. Rather than damaging the land or depriving academics of the opportunity to study their specimens, these collectors feel that they are safeguarding these Paleozoic treasures from harm, having possibly saved these fragile fossils from the disastrous fates of erosion or neglect.

In recent years, the lure of discovery has motivated some diggers to explore increasingly more remote fossil-laden localities—some currently on restricted government land and many not previously known to science. Some within the legislative arena now want to see such exploration severely limited, if not stopped entirely. With more new laws being passed on an annual basis, it is abundantly clear that the debate concerning the rules and regulations surrounding the worldwide collecting of fossils will only grow more heated, and derisive, as time marches on.

TRILOBITE VALUES

How to best value a specimen already in their collection, or one they may be thinking of adding to or removing from their Paleozoic holdings, is perhaps the most perplexing conundrum that routinely faces trilobite enthusiasts. That's true whether they are world-class veterans or relative beginners. It may not be as mind-numbing a consideration as contemplating the variables of Deep Time, or as intriguing as pondering the initial flowering of complex life on Earth, but it is a subject that every collector must face at one time or another. After all, a fossil isn't a commodity like gold or silver where a preordained international value can be affixed to it on a daily basis. Natural history items rarely, if ever, even enjoy the economic stability allotted to works of art, a market in which prices for individual paintings or sculptures notoriously rise and fall on the whims of an ever-fickle buying public.

Quite simply, where trilobites are concerned, values can be unabashedly arbitrary. Their prices are often based on little more than supply and demand. Here the time-worn axiom rings true: beauty, and any corresponding monetary worth, is very much in the eye of the beholder. For example, one of the hundreds of available *Asaphiscus wheeleri* specimens from Utah will invariably cost perhaps $50 to obtain, whereas the first example of a previously unknown ondontopleurid species from Russia holds the possibility of costing thousands more. But how much more is truly anyone's guess.

Numerous commercial websites, international auction houses, and accredited fossil dealers are only too happy to assign a lofty value to the au courant trilobite treasure in their possession. Too often, however, collectors are left wondering if the specimen they are being offered has been priced fairly or if they are just being treated as the latest sucker to walk down the primrose Paleozoic path with the appearance of deep pockets and shallow sensibilities.

Many hobbyists would like to believe that they collect fossils primarily out of personal curiosity rather than for any potential economic gain, but in recent years collecting trilobites has become something of a rich man's endeavor. At a certain level, these collections—in some cases containing literally thousands of specimens—clearly become investments as well as recreation. Yes, anyone can still go out in the field and with a bit of luck find a beautifully preserved trilobite in their local quarry, hillside, or streambed. And it's true that a nice, cabinet-sized assortment of common species can be sensibly assembled for $500 or less. However, to amass a comprehensive trilobite collection, featuring top-rated examples of unusual species gathered from major Paleozoic repositories around the globe, can easily run an enthusiast well over $100,000.

After all the paleontologically pertinent information is gathered, digested, and disseminated, what is the most logical means of determining

ASAPHISCUS WHEELERI (MEEK, 1873)

Middle Cambrian; Wheeler Formation; Utah, United States; 6.5 cm

This attractive, large, and well-preserved trilobite loses "points" for being among the most common North American species.

***ISOTELUS LATUS* (RAYMOND, 1913)**

Upper Ordovician; Cobourg Formation; Bowmanville, Ontario, Canada; Larger trilobite: 29.2 cm

This impressively sized double was *twice* sold off at major North American auction houses since its discovery in the 1970s. Despite representing a relatively common species, it wins "points" for its aesthetics and large dimensions.

the true monetary value of a trilobite? Is the price based solely on some marginally calculable combination of its appearance, its location, and its scientific gravitas? Do the number of freestanding spines adorning that latest Moroccan monstrosity dictate its retail appeal? For perhaps too many collectors and merchants that's apparently as good a criterion for judging value as any other. Some trilobite prices have recently risen to astronomical heights due to an intriguing mix of unexpected circumstances. These include the fact that during the initial decades of the twenty-first century

startling numbers of new species have begun to emerge from previously unknown localities. Many of these examples, including trilobites found in such distant destinations as Siberia, South Australia, and Northern Greenland, require excessive time, effort, and financial output simply to reach. All corresponding costs are naturally transferred to whomever eventually purchases a specimen resulting from these enterprising expeditions.

In addition, a good number of these exotic examples—especially those hailing from trilobite hotbeds such as Morocco and Russia—have benefited from dramatically improved preparation techniques, making them more aesthetically pleasing, scientifically revealing, and unapologetically expensive. Trilobite prices have also been directly affected by increased marketing opportunities provided by the internet, fossil trade shows, and international natural history auctions—all of which have helped introduce an often-affluent new demographic to the age-old wonders of these amazing arthropods.

"Trilobite collectors can be a strange group," said Virginia-based enthusiast Gregory Heimlich. "Some will only hold on to material that they personally dig from the ground. Others want a very diverse collection that features the best specimens from around the world. And while there are some unscrupulous sources around, the internet has made it relatively easy to cross-check both the prices and the availability of certain trilobite species."

As a point of reference, one of the world's leading trilobite collectors, who admits to having more than 3,500 complete specimens within his impressive Paleozoic stockpile, routinely uses a self-created 30-point rating system to place some sort of relative value on his various arthropod assets. He grades each trilobite from 1 (lowest) to 10 (highest) in three comparative categories: size, rarity, and quality. He admits to being a tough grader, with the slightest flaw on the trilobite's calcite-covered carapace (even if that blemish occurred during the creature's life, some half a billion years ago) possibly leading to the loss of points.

Most trilobite fossils are relatively small—under 5 centimeters in length—so size does often matter when it comes to assessing their perceived value. Specimens 8 centimeters or more stand a significantly better chance of earning a score of 10 than less robust examples, regardless of their corresponding appearance. Similarly, a particularly distinctive bug, or perhaps even a one-of-a-kind trilobite treasure, will tally higher than a relatively common species, no matter how exotic or pristine the latter may be. Predictably, a trilobite's taxonomic order also often plays a key factor in this rather random numbers game. Lichids, for example, almost invariably attain a higher point total (and subsequent value) than ordinary phacopids, strictly due to their comparative rarity and corresponding desirability to collectors around the world.

Of the tens of thousands of trilobites this collector has had pass through his fossil-fondling fingers over the past three decades, he insists that he's encountered perhaps no more than two dozen specimens that have scored a perfect 30—invariably they are rare or unique bugs from unusual locales that are perfectly preserved and impressive in size. He is the first to admit that such a rating system is subjective and the antithesis of foolproof. Despite its inherent limitations, his method at least provides a basic structure from which just about any serious collector with a keen eye, a cool sense, and a comprehensive knowledge of the trilobite market can gain a modicum of perspective regarding the value of their fossilized holdings.

In today's affluent global market, many of these prized 30 specimens can easily fetch $10,000 or more from high-end collectors in the United States, Japan, or Europe. Correspondingly, a seller can figure to lose roughly $2,000 a point until the trilobite in question drops to an overall score of 26. Beneath that number, values diminish even

MESONACIS SP.
Lower Cambrian; Rosella Formation, Atan Group; British Columbia, Canada; 8.1 cm

This striking specimen is similar to a species now found 3,000 miles away in the Lower Cambrian strata of Vermont.

more quickly, and by the time you consider a specimen that scores a solid 20, a seller would probably need to perform a sprightly song and dance upon a nearby tabletop to receive more than $100 for it.

"It used to be that there was an unspoken $1,000 threshold on trilobite prices," said Martin Shugar, a field associate with the American Museum of Natural History. "That was true up until the early 1990s when markets like Russia and Morocco opened up and preparation techniques turned many trilobites into true works of art. Now, for better or worse, the

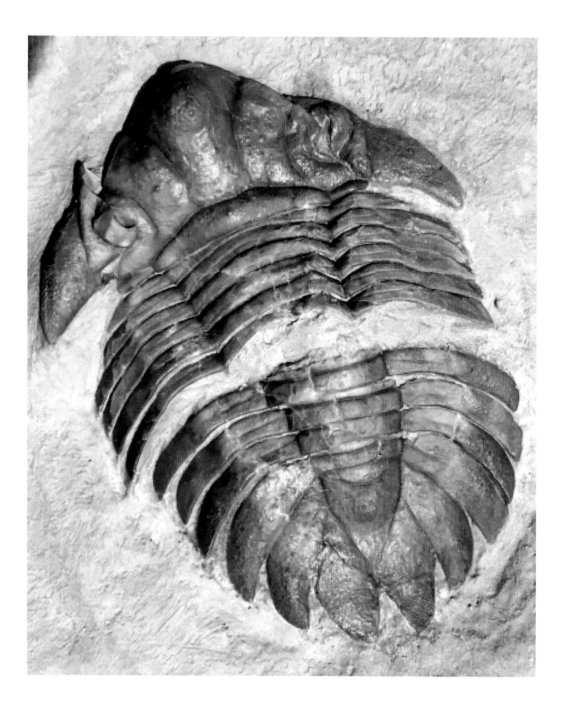

AMPHILICHAS OTTAWAENSIS (FOERSTE, 1919)

Upper Ordovician; Bobcaygeon Formation; Ontario, Canada; 6.1 cm

The break in this trilobite's thorax indicates that it is a molt—the discarded outer carapace a trilobite sheds a number of times each year to expedite its growth.

sky is the limit. But despite their recent increase in both price and value, I think it's fair to say that great trilobites have always been sought after."

These ancient arthropods have been viewed as prized collectibles for a very long time. Hand-drilled trilobites apparently worn as amulets have been uncovered in European archaeological sites dating back over 15,000 years. As early as the tenth century, Chinese homes commonly displayed trilobite specimens as honored works of natural art. Three hundred years ago, Native American tribes were known to carry small trilobite fossils in their medicine pouches, regarding them as a sacred talisman designed to ward off the powers of their enemies. It was probably the mid-nineteenth century before the world witnessed the first true commercialization of these intriguing invertebrates. At that time, small shops in both England and what was then Bohemia began to market the "frozen locusts" found in the nearby hills to locals or visitors who showed interest in them. However, what might have cost the equivalent of a few dollars then has now grown to become part of a lucrative and still expanding international industry.

Today both private collectors and public institutions annually spend extravagant amounts on the acquisition of spectacular trilobite specimens hailing from all parts of the globe. Just a few years ago, for example, one prominent North American museum reportedly spent over $35,000 on a large Silurian *Dalmanites limulurus* plate destined to be the centerpiece of a new fossil display, and at roughly the same time, a noted Asian facility was rumored to have laid down more than $200,000 to acquire a major private collection featuring more than 4,000 specimens. One of the primary reasons trilobite prices have risen so dramatically in recent years can be directly attributed to their increased availability on the global market. Any specimen, at any time, can now be placed on an international forum such as eBay, where prospective buyers from Anaheim to Antwerp can bid against one another to their heart's content. The net result of such auction action has prompted more trilobite material than ever to appear on these outlets, with prices for particularly desirable specimens spiraling continually upward.

"It used to be that major fossil transactions were done under the table at trade shows," said Bill Barker, a leading commercial fossil dealer. "Now they're conducted in front of the world on the web. It's brought more 'players' into every significant deal, and prices have been impacted accordingly."

This internet-based revolution has led to a quantum shift in hobbyists' access to trilobites, as well as in their willingness to pay often outrageous prices to land these Paleozoic treasures. Want a flawless double *Asaphus latus* specimen from Russia? The Bonhams.com auction site has one for $2,000. Will your life never be complete without a spiny *Drotops megalomanicus* trilobite from Morocco? There's one over on eBay that can be had for a mere $1,500. (Oh, sorry, it *just* sold.) How about a diminutive, but incredibly rare *Sphaerocoryphe robusta* from the famed Walcott/ Rust trilobite quarry in New York State? It's yours for $5,000 over at Fossilera.com if you click fast enough! Indeed, the internet has not only propelled trilobite renown to previously unimagined levels, it's also helped establish an accepted, if often inflated, value for many unusual trilobite species.

Scores of fossil-oriented outlets exist across the internet—many run by dealers and collectors from all over the world—with each site expressly designed to offer up tasty trilobite morsels to the seemingly insatiable hobbyist horde. From $10 Moroccan *Calymene* specimens to $10,000 *Isotelus gigas* plates from Ontario, over the past two decades the internet has steadily evolved into a heady cross between flea market and high-end auction house. It is geared for presenting and selling some of the most incredible trilobite finds ever offered for public consumption—at prices that continually manage to boggle the mind. The inherent nature of internet sales seems to spur the competitive juices of many trilobite collectors, motivating them to surpass the boundaries of their previously ordained budget to succeed in their somewhat quixotic quest. But the internet has also had the beneficial effect of taking previously unknown and unobtainable species from such distant lands as China, Bolivia, and Australia and transforming them into examples that can be acquired for a

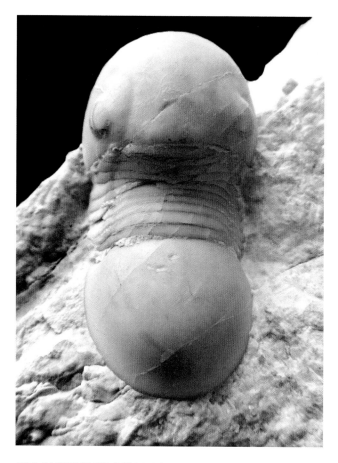

BUMASTOIDES BILLINGSI (RAYMOND AND NARRAWAY, 1908)

Middle Ordovician; Prosser Formation; Cherry Grove, Minnesota, United States; 4.5 cm

This attractive specimen reflects the characteristic golden color exhibited by the trilobites found in this formation.

surprisingly reasonable price (usually under $300) with just the quick click of a mouse.

"There is now a near-constant flow of trilobite material from places like China and the Czech Republic on the internet," said Pavel Dvorak, a Czech trilobite specialist. "Those specimens usually aren't perfect or particularly attractive, though some are. But that kind of material was hardly ever available to the collector just a few years ago."

The internet has opened incredible new vistas of opportunity for a vast number of collectors, including those looking for rare fossils costing thousands of dollars. Many long-time followers of the trilobite trade have been astonished by the voracity of the collecting community, as well as by the amount of money serious hobbyists will spend—often without seeing more than a poorly lit digital photo of the trilobite in question prior to completing their purchase.

With the astronomical prices top-grade trilobite fossils now routinely fetch from enthusiasts around the world, and with the collecting community itself growing at a healthy rate, you might assume that the values attached to at least some of these primal organisms would continue to escalate in the years ahead. But can hobbyists feel relatively safe that their bankbook-bending expenditure on an idiosyncratic item such as a trilobite represents a sound economic investment? Unfortunately, the answer to that one is probably no. Even prestigious publications such as *Forbes* magazine and the *New York Times* have recently acknowledged the occasionally risky financial outlay required for trilobite acquisitions. After all, there is no DeBeers-like cartel overseeing the trilobite trade, controlling where diggers can search for these Paleozoic treasures, then monitoring the flow of material to market and reacting to any subsequent rise or fall in sales action. Quite simply, there is no convenient or sensible means of ensuring that trilobite prices stay fair and values stay true.

As more monetarily motivated diggers begin exploring deeper and deeper into the Earth's sedimentary recesses, the chances of finding a bonanza of previously unknown trilobite species or perhaps additional examples of hitherto unique ones continue to skyrocket. Thus, a specimen that rates as a rare 30 today may turn into something comparatively common tomorrow. That's an unpleasant phenomenon some collectors have experienced time and time again over the past decade, especially when dealing with top-dollar trilobite material emerging from Morocco and Russia. One year they're being offered a first of its kind Cambrian rarity, and just 12 months later a table full of such specimens are selling for less than half of the previous year's asking price!

With that in mind, let's return in a rather roundabout manner to the original subject—how best to judge a trilobite's monetary value. Aside from relying on rather arbitrary 30-point rating systems or perusing self-serving internet price guides, there is no obvious manner of calculating either the wisdom or the stability of such trilobite-centric investments. When all the various elements are weighed, perhaps the most effective means of measuring the value of these increasingly coveted primeval relics is by asking a very basic question: What is that particular trilobite worth to the one who owns it? The market for natural history items will continue to rise and fall, but the true value of your half-billion-year-old treasure rests solely within your heart rather than your pocketbook.

"You buy a trilobite and own a trilobite because you like that trilobite," said Dvorak. "If you are doing it for potential financial gain, you're playing a very dangerous game. Collecting fossils is a wonderful hobby that can expand your imagination. Just don't expect it to also expand your wallet."

ACQUIRING A TRILOBITE

Amid this book's focus on trilobite locations, trilobite morphology, trilobite values and trilobite history, we have not yet addressed a very basic question: How does one best acquire a collection-quality trilobite specimen? Back in the "old days"—say, 1990—to procure a world-class trilobite you had to journey out into the field. There you would be forced to brave the unpredictable forces of nature as you battled against the elements and perhaps even your digging partners in your attempts to pry a prize example out of the sedimentary soils. And, come to think of it, that may still represent the

Ultimate Trilobite Challenge—finding your own display-worthy specimen through hand-blistering hard work. However, unless you happen to live near an established trilobite-bearing outcrop, or you're willing to risk time, energy, and expense exploring for your own special dig site (and you better make sure it is not protected by any freshly enacted government restrictions before you break any rocks), odds are that your efforts are going to produce little or nothing worth writing home about.

If you don't particularly want to either freeze or fry in the field, you can visit one of the major fossil shows that take place annually in Tucson, Munich, and Tokyo. It is there that fossil dealers from both near and far gather to present a veritable Paleozoic buffet designed expressly to entice those with varying degrees of knowledge, desire, and disposable income to part with their hard-earned cash in exchange for a top-tier trilobite specimen. And if neither traveling into the field nor attending a trade show particularly lights your fossiliferous fires, don't overlook the resources of the trilobite-collecting community itself. Enthusiasts often enjoy interacting with their fellow fossil fanatics, whether by phone, on Facebook, or in face-to-face encounters. These experiences provide an opportunity for you to view prime specimens, swap gossip and information, and trade duplicate examples. There is perhaps no faster or more entertaining means of expanding a collection, or acquiring that one Dream Bug, than by dealing directly with those who share your paleontological passions.

However, if none of these activities serve to fully scratch your trilobite itch, there is one more place to turn. No longer do you have to plan a journey by foot, car, train, or plane to see and acquire the latest and greatest in trilobite treasures. Indeed, all you need do is sit in a cushy chair in the comfort of your own home and flip on your computer to find an entire array of incredible fossil-collecting opportunities for both the novice and the advanced

HOPLOLICHOIDES CONICOTUBERCULATUS (NIESZKOWSKI, 1859)

Upper Ordovician (Sanbian); Kukruse Regional Stage; Alekseevka Quarry; St. Petersburg region, Russia; 5.9 cm

The ivory and red color contrast of this spinose specimen reflects the stunning preservation found on trilobites being drawn from a recently opened quarry in western Russia.

consumer of natural history goods. Dozens of fossil-oriented websites feature eye-catching trilobite specimens from throughout the world at price points ranging from pocket change to profoundly big-ticket. Whether your intent is to assemble your own museum-worthy trilobite collection or merely to obtain a single top-grade example, there

METOPOLICHAS PATRIARCHUS (WYATT-EDGELL, 1866)

Lower Ordovician; Llanvirnian Stage; Stapeley Shales Member, Middleton Formation boundary; Gravelsbank, Hope Valley, Shropshire, England, UK; 6.6 cm

Unusual species such as this one are best acquired directly from old collections. Rarely do these distinctive trilobites appear on the open market even at the most renowned fossil shows.

are many arthropod-lined avenues to pursue. With any luck, your actions will result in you becoming the proud owner of a unique piece of our planet's distant past.

CURATING A TRILOBITE COLLECTION

You've already done the hard work and assembled a sizable trilobite collection containing dozens (if not hundreds or even thousands) of specimens from all the corners of the globe and all the periods of Deep Time. They line cabinets in your home, fill shelves in your office, and overflow boxes in your upstairs closet. What the heck do you do next? Perhaps the moment has come to make better sense of your voluminous holdings by properly curating your collection. And thanks to the prodigious amount of information available on the internet—ranging from enlightening museum-sponsored sites where a variety or trilobite-oriented papers, photos, and journals can be found to amazingly detailed amateur pages—collectors have access to a world-class compendium of essential trilobite data literally at their fingertips.

No matter its place of origin, its age, its rarity, its condition, or the degree of academic attention it may have already garnered, every collection-worthy trilobite should be provided with an accompanying identification number and label that can be quickly and easily cross-referenced with a list stored conveniently in a notebook, on a series of index cards, or in a well-marked computer file.

Thus, whenever anyone comes across a specimen, whether it is "*Belenopyge balliviani* (Koslowski, 1923), Lower Devonian, Belen Formation, La Paz, Bolivia," or "*Bolbolenellus brevispinus* Palmer, 1998, Lower Cambrian, Pioche Formation, Nevada, USA," that trilobite's vital info will be immediately apparent, shedding light not only on its scientific history but also on its role within the Paleozoic's evolutionary puzzle.

CERAURUS PLATTINENSIS FOERSTE, 1920

Upper Ordovician; Lower Verulam Formation; Ontario, Canada; 6.8 cm

Sometimes a near-perfect specimen such as this can be damaged during its extraction from the surrounding matrix.

Some in both the scientific and collecting communities will eagerly inform you that without proper documentation and labeling—identifying each specimen as to its genus, species, author (the person who first described the trilobite in literature), location, formation, and age—your prized

BOLBOLENELLUS BREVISPINUS PALMER, 1998

Lower Cambrian, Series 2, upper Dyeran; Pioche Shale, Combined Metals Member; Panaca, Nevada, United States; 3.4 cm

These small trilobites feature an extremely thin outer shell, making them difficult to find and prepare.

trilobite is little more than a primeval curio, a Paleozoic paperweight. Those on the extreme edge of this argument may even state that if your trilobite wasn't collected under the administrative auspices and direct supervision of a trained professional—someone not only able to note a field specimen's geographical alignment and proper sedimentary deposition but also the exact GPS coordinates of its discovery—that trilobite has lost all of its paleontological gravitas. To that I loudly and clearly say, "bull spit." Every trilobite specimen—no matter where it may have been found or under what conditions it was collected—deserves some degree of respect and recognition. If such stringent requirements were placed on the seemingly endless assortment of previously unknown, usually uncharted, and so-far undescribed species currently emerging from the Paleozoic rocks of Morocco, it could be argued that every such trilobite is devoid of scientific value.

In truth, it would be difficult, if not downright impossible, to properly identify and label many of these North African specimens until a great deal more research is done on them. That notion should not make such trilobites any less collectible, or lower either their fossiliferous appeal or their intellectual importance. A temporary label, perhaps indicating the specimen's presumed genus, geological age, and country of origin, can suffice until additional information becomes available.

Despite such minor distractions, the proper curation of their collection ranks among the most pleasurable aspects of their hobby for a multitude of trilobite devotees. In many cases, this daunting yet eminently fulfilling process provides enthusiasts with a true sense of academic accomplishment. Thanks to their efforts, some 500 million years after their chosen trilobite met its untimely demise, it now possesses a restored level of identity that will enable it to be studied, admired, and enjoyed for many eons to come.

BEHIND THE SCENES AT THE AMERICAN MUSEUM OF NATURAL HISTORY

Filled as they are with artifacts, antiquities, and fossils from all corners of the planet and all aspects of time, it seems prosaic to say that natural history museums rank among the world's most captivating places. Indeed, such institutions often serve as *the* premier tourist destination within their respective city or country, repositories capable of annually attracting hundreds of thousands—if not millions—of visitors, all eager to enjoy that establishment's featured exhibits, colorful dioramas, and enlightening geological displays.

With so much to see, and so much to do, a trip to any major natural history museum is always a memorable experience. For some, it may well represent their sole opportunity to directly encounter the anthropological, archaeological, and paleontological curios that have fueled their fantasies since childhood. Where else but within museum sanctuaries can one find roof-scraping dinosaur skeletons vying for attention with meteorites the size of small cars and wall displays featuring a world-spanning array of *Paradoxides* trilobites dramatically revealing the dynamics of plate tectonics?

A museum is a place where our imaginations run wild and our dreams come alive, and few such institutions have ever come close to matching the "dream-machine" panache of New York's famed American Museum of Natural History. It is within the AMNH's cathedral-like galleries and dramatically darkened alcoves that generations of New Yorkers have grown up gawking at cases filled with menacing *Triceratops* horns and delicate pterosaur wings. Every year more than five million visitors travel from both near and far to pass through the museum's formidable brass doors, with each hoping to experience something they have never experienced before.

Much like an Arctic iceberg, where only a small fraction of its frozen mass protrudes above the surrounding sea line, what the public views on display within the AMNH's 190,000 square meters of busily aligned shelves and brightly lit cases represents less than 1 percent of what that facility has sequestered in its back rooms and hidden laboratories. There, amid a dizzying labyrinth of interconnecting passages and crowded hallways befitting any Hollywood-inspired *Night at the Museum* vision, row upon row of towering cabinets house everything from unprepared sauropod skulls to intricate trilobite mass mortality slabs.

Clearly, any visit to the American Museum of Natural History carries its own unique allure. But that's especially true when one is granted a private peek beneath the museum's time-honored façade to witness what transpires behind the scenes and to see treasures that have been hidden away—often for decades—from the prying eyes of the public. Even within such a hallowed institution, however, things aren't always exactly as they may initially appear.

It was late a Friday afternoon, a traditionally quiet time at the AMNH. The roving packs of boisterous schoolchildren and tightly clustered pods of wide eyed out of towners who only hours earlier had crammed the establishment's massive dinosaur halls to stare *Tyrannosaurus rex* square in the face had long since headed home or retreated to the refuge of nearby hotels. While rush hour traffic rolled past the AMNH's Upper West Side locale in an ear-assaulting cacophony of horn blasts, inside

KOOTENIA YOUNGORUM **ROBISON AND BABCOCK, 2011**
Middle Cambrian, Series 3, Stage 5; Upper Spence Shale, Glossopleura Zone; Wellsville Mountain, Utah, United States; largest trilobite: 6 cm

This is an engaging quartet featuring a rare trilobite species that was scientifically described in 2011.

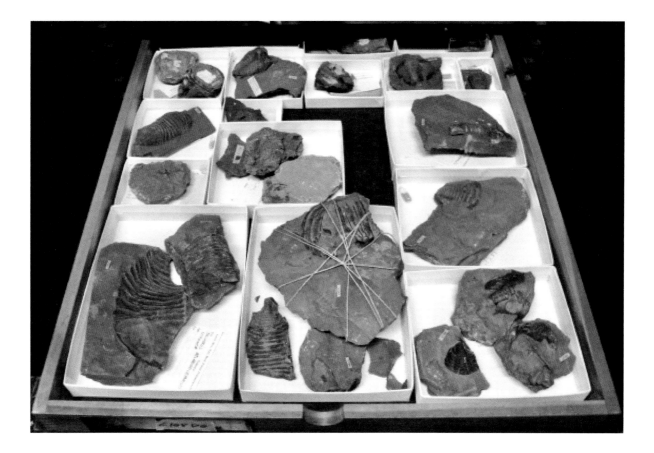

A TYPICAL MUSEUM TRILOBITE DRAWER.

the museum's imposing structure a library-like pall provided a stark and compelling contrast. It was an ideal time to visit this renowned Big Apple landmark—a moment when one could focus on a specific point of paleontological interest without a great deal of fuss, muss, or bother.

Shortly before 4 p.m., our small group was greeted in the museum's cavernous Seventy-Seventh Street Grand Gallery by a staff member who had been informed beforehand that our primary interest would be to view the AMNH's trilobite holdings. We had agreed to meet near a recently installed plexiglass case that featured a small selection of top-of-the-line trilobite specimens from Russia, Morocco, and North America, including superlative examples of *Asaphus kowalewski*, *Dicranurus monstrosus*, and *Spathacalymene nasuta*.

Immediately after meeting, our amiable host guided us through the public halls to show off the few trilobites that were on display—a small yet attractive Ordovician *Isotelus iowensis* in the Hall of Ocean Life, a Cambrian *Olenellus fremonti* located along the spiral Cosmic Pathway that provided egress from the nearby planetarium, and a nicely prepared Czech *Paradoxides gracilis* featured in the Hall of Planet

Earth concourse. Then we came to the time-tested Hall of New York State Environment—which has remained essentially unchanged since its inception in the 1940s—and expectations grew.

After all, the museum was renowned for housing a sizable portion of the original James Hall collection of Rochester Shale trilobite material found in upstate New York during the mid-1800s, and advance word was that a selection of those Silurian specimens was on display here. Alas, we were soon confronted by a few poorly preserved examples of *Dalmanites limulurus* and *Trimerus delphinocephalus* that some 80 years earlier had been placed in a large case that now appeared to feature more Holocene-age dust than Paleozoic Era fossils.

Our group was then escorted upstairs to view the remainder of the Rochester Shale holdings in the "staff only" confines of the museum's expansive fifth floor archives. Amid rows of 3-meter-high metal cases, some featuring hand inscribed labels reading "Trilobita: *Arctinurus*," our anticipation of pending paleontological euphoria once again began to rise. To our delight, there were plenty of trilobites to see—most apparently unearthed and added to the museum's collections nearly a century ago. But as we were soon to learn, although many possessed unequivocal scientific gravitas, along with a certain vintage charm, few exhibited the aesthetic "wow" factor that we, perhaps unrealistically, had been seeking. In truth, much of the museum's Rochester Shale trilobite assemblage seemed to be comprised of disarticulated and ventral heads or tails, with complete specimens clearly being the exception rather than the rule.

No matter how hard we looked, and how many drawers we opened in our rapidly expanding, late afternoon trilobite search, it was nearly impossible for anyone in our group to find a specimen that would have seemed particularly out of place in a friend's midsized collection. Indeed, the vast preponderance of the museum's trilobite holdings

ELLIPTOCEPHALA SP.

Lower Cambrian, Series 2, Dyeran Regional Stage; Middle Member of the Poleta Formation; Montezuma Mountains, Esmeralda County, Nevada, United States; 10.3 cm

This 520-million-year-old example beautifully displays the advanced evolutionary state that trilobites had already achieved soon after appearing in the primal seas.

could have perhaps best been described as "study grade" material . . . with hardly a display-worthy specimen in sight. When our genial guide brought attention to a complete example of the common Czech species *Ellipsocephalus hoffi*, one member of our party only half-jokingly suggested that a nineteenth-century Prague opera ticket—the reverse of which served as the specimen's handwritten identification label—was of more potential interest than the trilobite itself.

Our small group left the AMNH that day wondering what exactly transpired within the Ivory Towers of many natural history museums. Amid a flurry of opinions offered during a quick stop at a nearby pizza joint, one key question seemed to dominate our conversation: Why would an institution that seemed to bask on the cutting edge of dinosaur research and discovery be so seemingly content to take a back seat when it came to compiling new material from as compelling a fossil class as trilobites?

We all realized that in these unpredictable economic times, perhaps the funds needed to acquire exciting new trilobite specimens were not as readily available to major institutions as we might have initially surmised. We were also aware that finding staff members with both the background and expertise needed to serve as prime acquisitors of such fossil material could prove to be far from an easy task. Our group was pleasantly surprised to learn only a few months after our visit that the AMNH had hired a new curator, Melanie Hopkins, whose field of study was, in fact, trilobites. Even more encouraging, one of her primary missions was reportedly to increase both the breadth and

scope of the museum's display-worthy Paleozoic holdings, a task she has already successfully begun to tackle—partly in conjunction with the 2022 opening of the facility's expansive Gilder Center for Science, Education & Innovation.

Still, our rather eye-opening experience at the AMNH got our group to thinking: What responsibility—if any—does a major institution carry when it comes to obtaining the latest and greatest in fossil treasures? And if a museum happens to make such acquisitions, what obligation does it then have to put those specimens on prominent display?

Apparently within the mindset of many natural history museum hierarchies, a belief still lingers that direct viewing of their most important fossils should be reserved strictly for those involved with scientific study and research. Such an attitude is understandable and even somewhat practical, at least to a certain extent. But just imagine if leading art museums were to follow a similar approach, sequestering their prized Rembrandt, Rodin, or Picasso behind closed doors and presenting a well-designed print or cleverly crafted copy for public consumption—a somewhat deceptive "trick" many natural history establishments utilize with everything from replica dinosaur bones to resin-based *Archaeopteryx* plates.

In stark contrast, some more au courant facilities, such as the Houston Museum of Natural Science—which features not only the best public display of trilobites in North America but also no less than *three* complete and genuine *T. rex* skeletons on permanent exhibit—seem almost exclusively geared to showcasing their latest acquisitions as high-profile tourist attractions. For better or worse, these institutions have chosen to place a sizable portion of their assumed academic responsibilities in a secondary position to pure shock and awe showmanship.

The sense is that for world-renowned establishments, such as the American Museum of Natural

(OPPOSITE PAGE, TOP LEFT) **LONGIANDA TERMIERI HUPÉ, 1953**

Lower Cambrian; Issafen Formation, Zone Vi, Sectigena Zone; Issafen, Morocco; 13.4 cm

This is a long-recognized species—first described from disarticulated bits and pieces—that has only begun yielding complete fossilized examples in the last two decades.

(OPPOSITE PAGE, BOTTOM RIGHT) **ARCTINURUS BOLTONI (BIGSBY, 1825); STEGERHYNCHUS NEGLECTUM (BRACHIOPODS)**

Lower Silurian; Rochester Shale Formation; Caleb's Quarry; Middleport, New York, United States; 13.1 cm

This beautifully preserved specimen is covered in epibionts (brachiopods), which probably attached themselves to the trilobite's exoskeleton while it was still alive.

History, the Smithsonian Institution (which in 2019 completed a comprehensive remodeling of its fossil halls), or London's Natural History Museum, properly maintaining, let alone expanding, collections that have been amassed over decades, if not centuries, is an overwhelming task. Too often there are just not enough curators or able-bodied assistants available to provide a proper degree of academic attention to the tens of millions of often fragile specimens that already fill major museum storerooms.

In recent years, the perceived pressure of adding to their fossil collections has emerged as a major source of consternation and concern for many natural history establishments around the globe. With operating costs shooting through the roof, and the expense of promoting expeditions to faraway lands in search of new Paleozoic and Mesozoic wonders becoming more and more prohibitive, some museums have started to turn their backs on the pursuit of new acquisitions. Instead, these facilities have begun focusing inward, reexamining their already on-site holdings with the intent of uncovering previously overlooked, misidentified, and potentially significant paleontological gems. Other museums have begun to rely more heavily on the generosity of outside benefactors, who they hope will either donate desired specimens or contribute the funds needed to acquire such material. Somewhat ironically, this escalation in financial support hasn't necessarily made it easier for those benefiting institutions to obtain the fossils they most covet.

The expanding role of the internet along with the emergence of semiannual natural history auctions held by the likes of Sotheby's and Christie's in New York, Paris, and Los Angeles have allowed museum-worthy specimens to become increasingly more available to wealthy private collectors around the world. When combined with the expanded interest shown toward upscale fossil material by those who view these primal relics as either "natural art" or inflation-hedge "investments," it has all significantly upped the ante when it comes to the competitive nature of acquisition—rarely the strong suit of a museum curator.

For a variety of reasons, during the first two-plus decades of the twenty-first century, many natural history museums have grown increasingly lax when it comes to directing their attentions or their financial resources toward the procurement of new fossil material—including trilobites. The resultant paleontological paucity has begun to present an unexpected conundrum for institutional officials—a problem centered around a delicate balancing of updated acquisitions, collections, and displays that is not likely to be rectified any time soon. Indeed, many interested observers have begun to speculate that if events continue to evolve as they have in recent years some of these esteemed establishments may eventually run the risk of becoming as musty and out of date as their fast-aging exhibits.

We can all only hope that such a sad state of affairs never comes to fruition, for the very real concern is that the diminishment of any such institution may also signal the lessening of the empirical hopes, dreams, and aspirations that have fueled mankind's forward progress for millennia. Much like the old proverb that warns that for the want of a nail the kingdom was lost, it may well prove that for the want of a trilobite the museum was lost.

TOP TRILOBITE MUSEUMS

Even within the most prestigious natural history museums, trilobites rarely receive their proper degree of recognition and respect. Too often the fossilized remains of these ancient organisms are relegated to small cases in dark corners of back rooms as the oversized bones of dinosaurs and woolly mammoths vie for the saber cat's share of front lobby acclaim. But despite the less than stellar manner in which they are frequently presented to the public, a number of notable institutions around the globe have at least attempted to prioritize their display of trilobites. Here is a brief look at some of the World's Top Trilobite Museums.

Houston Museum of Natural Science, Texas: This relatively new and expansive facility in the American Southwest highlights what is perhaps the continent's largest and most impressive public trilobite exhibit, one that primarily focuses on the aesthetic presentation of exotic Moroccan and Russian species. More than 150 prime specimens, drawn from across the face of the planet and representing every Paleozoic period, can be viewed in a sequence of spacious, well-lit cases. A choice number of North American trilobites are also showcased, including *Olenellus fremonti* from California and *Gabriceraurus dentatus* from Ontario.

Royal Ontario Museum, Toronto: The ROM just happens to house the world's largest collection of Burgess Shale material, so it should come as no great surprise that an eye-popping display of that locale's hallowed Middle Cambrian fauna—featuring a formidable assembly of soft-bodied organisms—is on exhibit within this all-encompassing facility. The museum's Paleozoic presentation also includes such unique trilobites as an *Olenoides serratus* with soft-tissue preservation. Of particular note is perhaps the only semicomplete example of the famed "trilobite eater" *Anomalocaris canadensis*.

American Museum of Natural History, New York: This legendary Big Apple institution is certainly more renowned for its grand dinosaur galleries, which encircle the entire fourth floor, but there is a small, comprehensive trilobite display located in the museum's Seventy-Seventh Street lobby. Among the baker's dozen featured specimens—which are conveniently presented

OLENELLUS GETZI DUNBAR, 1925
Lower Cambrian; Kinzers Formation; Brubaker Quarry;
Lancaster, Pennsylvania, United States; 14.2 cm

The distinctive mineralized preservation of these trilobites occasionally display antennae. Unfortunately, none are seen on this large specimen.

in proper time sequence order—are a pair of spectacular Silurian offerings, *Spathacalymene nasuta* from Indiana and *Arctinurus boltoni* from New York. There is also a massive *Xenasaphus devexus* plate from the Ordovician outcrops that surround St. Petersburg, Russia.

Natural History Museum, London: What was known until 1992 as the British Museum

THALEOPS OVATA (CONRAD, 1843)

Upper Ordovician; Black River Group, Leray Formation; Quebec, Canada; 3 cm

This attractive specimen lies near a coral fragment that provides an indication of the tropical climate in which it existed.

(Natural History) presents a splendid assemblage of indigenous trilobites, most adorning a series of sturdy, prime location cases. There is a definite—and understandable—emphasis placed on the magnificently preserved Silurian specimens from nearby Dudley. But among the other featured trilobites to be found in this dramatic Romanesque structure (first opened in 1881) are an intriguing selection of *Ogygiocarella* examples hailing from throughout the British Isles.

Smithsonian Institution, Washington, D.C.: Recently renovated, the Paleozoic exhibit in the Smithsonian features some of the most impressive trilobite specimens to be found anywhere in the world. Among the key items are a rare *Bathynotus holopygus* from Vermont and an equally unusual *Trimerus vanuxemi* from West Virginia. There is also the only known, complete *Apianurus sp.* from the hallowed Walcott/Rust quarry of New York State. However, as nice as the new displays are, many visitors still miss the old Paleozoic cases that included a large *Isotelus maximus* from Ohio and a complete 12-centimeter *Dikelocephalus minnesotensis* from Wisconsin.

Hunterian Museum & Art Gallery, Glasgow: The beautifully preserved Ordovician trilobites of Girvan, Scotland, have long been admired and studied from London to Los Angeles. No museum contains a more comprehensive grouping of this distinctive material than the University of Glasgow's own Hunterian Museum. Enriched over decades by the donations of many notable local collectors, this institution both houses and displays a rich variety of the area's outstanding Paleozoic fauna, including stellar examples of such trilobites as *Paracybeloides girvanensis* and *Uripes maccullochi*.

Field Museum, Chicago: A centrally positioned circular structure that somewhat resembles a glass-encased UFO presents the best trilobites held in this famed midwestern museum. Among the featured specimens are a smattering of relatively common Russian and Moroccan species, along with a few key North American examples, including *Olenoides nevadensis* and *Homotelus bromidensis*. The Field also spotlights a notable display of *Paradoxides davidis* from the famed deposits of Manuels River, Newfoundland, all donated by the late Riccardo Levi-Setti.

Czech National Museum, Prague: Housing a preponderance of the legendary Czech trilobite collection gathered in the nineteenth century by Joachim Barrande, prior to its recent (and seemingly never-ending) renovation this facility presented one of the most comprehensive regional displays to be found in Europe. Now fewer specimens are seen but under better lighting conditions and with less apparent, decades old dust. Highlights include breathtaking examples of locally discovered *Bumastus hornyi* and *Dalmanitina socialis*, both of which have long graced the pages of the planet's leading scientific publications.

Museum of Comparative Zoology, Boston: The few trilobites on public view in this medium-sized facility located on the campus of Harvard University are nothing to write home about. But in its back rooms, the MCZ can boast of possessing one of the premier museum trilobite collections in the world. In addition to an outstanding arthropod array from New York's Walcott/Rust quarry (many acquired directly from Walcott himself), it features the most complete European trilobite assemblage to be found in a North American institution—including an extensive Barrande-sourced hoard of Czech material—as well as a unique assortment of the locally discovered *Paradoxides harlani* from Braintree, Massachusetts.

Back to the Past Museum, Cancun: Billed as "The World's First Trilobite Exclusive Museum," this impressive display in Cancun, Mexico—representing the arthropod-obsessed efforts of a single, private collector—is part of a trilobite-themed luxury resort. There is even a morphologically accurate, trilobite-shaped mosaic adorning the pool bottom! Hundreds of superlative specimens that cover all ages from Lower Cambrian to Permian are on exhibit, with highlights including a variety of rare *Bristolia* species from Nevada.

PREP WORKERS: TRILOBITE TRANSFORMERS

It looked like a rock. It felt like a rock. On closer inspection, it even exuded the distinctly earthy aroma that made it smell like a rock. The piece in question was expectedly heavy and rather nondescript, a predominantly dark-gray mass featuring a blend of smooth and jagged edges that provided this particular slab of sedimentary stone with the appearance of a somewhat squashed softball. To most of the people meandering through the

MICROPARIA NUDA WHITTARD, 1961
Lower Ordovician, Llanvirnian Stage; Shelve Formation, Stapeley
Volcanic Member; Shropshire, England, UK; 2.6 cm

While alive, these small trilobites featured huge, wraparound
eyes that provided an unparalleled view of the seafloor beneath
them. Unfortunately, those eyes are not preserved on this
specimen.

Denver Fossil and Mineral Show that mid-September afternoon, the item was nothing more than what it appeared to be. It was, indeed, a rock! Yet this specific chunk of Devonian limestone had drawn the interest of two intrepid individuals who had begun studying it with thinly veiled enthusiasm. At first they examined the specimen with their naked eyes, holding it mere centimeters from their faces while rapidly commenting back and forth throughout the process. Then one of the two decided to pull out a pair of well-worn glasses from his jacket pocket, somewhat haphazardly placing

them on the bridge of his nose, and never missing a beat in the running discourse.

The rock was carefully rotated and studied from every available angle. Not a fraction of its craggy surface missed detailed inspection. Moments later a hand lens was borrowed from a nearby acquaintance and the examination moved on to a more intense level of observation. The two proceeded to turn the rock over and over, angling it this way and that, making sure that the best possible light could shine on certain key nooks and crannies. One even impulsively used a bit of saliva to wet down a spot of special interest so it contrasted more prominently upon the stone surface. It was then that the finger pointing began, with three diminutive marks on the rock's mottled front side drawing the pair's most ardent attention. Strolling show visitors began to be pulled into the scene by both the excited dialogue taking place and the ever-wilder gesticulations that seemed to increase in direct accord with the two's animated discussion.

What were those guys looking at? Were there flakes of gold buried in that weathered matrix, or possibly some sort of precious gem? Maybe it was a Martian meteorite. Actually, it was nothing of the sort. But to the eyes of our paleontologically inclined rockhounds, what they had stumbled upon was perhaps even more exciting than a section of some distant planet, or even a diamond in the rough. With their highly trained senses, these fossil preparators had noted a series of small, circular marks permeating the sedimentary rock surface, none bigger than the graphite core of a no. 2 pencil. What they had found housed within this 420-million-year-old stone was the barely there cross-section of a rare genus of spiny Moroccan trilobite called *Radiaspis*. As both were well-aware, the best trilobites in the world were usually prepared from specimens discovered in cross-section, where there was far less chance that fragile axial spines or detailed eye facets could suffer damage during initial extraction.

Once they had negotiated an acceptable purchase price with the eager North African merchant who was offering the piece for sale, neither could wait to get this promising fossil-bearing stone back to their lab. After being subjected to a 20-hour or longer preparation process, they hoped it would emerge transformed, with the trilobite treasure hidden within the rough limestone matrix revealed in all its Paleozoic glory.

"Some people see a beautiful trilobite, and they assume that it was originally found that way," said Zarko Ljuboja, an Ohio-based fossil preparator whose specialty is trilobites. "Occasionally, especially with some Lower Cambrian trilobites, that can happen. The shale is split, and the entire specimen is revealed. But most of the time the trilobite is either buried in the rock with only small pieces of shell showing, or you see them in cross-section, where, if you're lucky, you'll see the thin outline of the trilobite's profile. It usually takes a lot of imagination, along with a fair amount of skill, to turn that into something special."

Michelangelo once explained his approach to sculpture by stating that he merely freed already existing pieces of art from their surrounding stone coverings. Certainly, that level of aesthetic intuition provides

REMOPLEURIDES NANUS ELONGATUS (SCHMIDT, 1894)

Upper Ordovician, Caradoc Series; Lowest Kukruse Regional Stage; Viivikonna Formation; Kingisepp Quarry, St. Petersburg region, Russia; 2 cm

Middle Ordovician trilobites from Russia had become standard fossil fare by the second decade of the twenty-first century, but Upper Ordovician specimens such as this one subsequently began to amaze academics and collectors worldwide.

key insight into the depth and breadth of his genius. But while the Renaissance master's words may have reflected the unique thought process that propelled the creation of his renowned works, when the subject shifts to those who prepare trilobites, such a method of extraction is *exactly* how it's done. Perhaps those who deal with fossils don't possess quite the inherent level of artistic elan exhibited by the famed Michelangelo, but when all is said and done, it is they who truly free existing forms from their encasing stone matrices.

Yet for all their apparent skill, the deft work done by trilobite preparators has only become properly acknowledged during relatively recent times. Particularly over the past three-plus decades, things have changed radically within the trilo-prep world. If we venture back to the mid-1980s, fossil preparation was still basically an arduous, often haphazard process in which acid baths, wire brushes, and handheld dental tools frequently left the trilobite as little more than a bruised and battered remnant. By the dawning of the twenty-first century, however, prep work had evolved into a state-of-the-art procedure employing an exotic assortment of pneumatic scribes, air-abrasive machines, and powerful vacuum ventilators. In the proper hands, these devices were often capable of miraculously transforming even the most derelict half-billion-year-old trilobite into a nearly flawless Paleozoic prize guaranteed to garner admiration from both the academic and collecting communities.

Many of these "next-generation" trilobites now feature delicate freestanding spines—some no thicker than a strand of angel hair pasta—and

ELLIPTOCEPHALA PRAENUNTIUS (COWIE, 1968)
Lower Cambrian, Series 2, lower Dyeran; Poleta Formation; Montezuma Range, Esmeralda County, Goldfield, Nevada, United States; 6.4 cm

This unusual species popped out of the half-billion-year-old rock as a near-perfect positive/negative split.

amazingly complex compound eyes. Such peculiar yet appealing characteristics serve to make these primal examples both a joy to observe and a pleasure to study . . . if, perhaps, an occasional pain in the pygidium to prep. Thanks to the myriad morphological features revealed through recent advances in preparation techniques, trilobites have garnered a new and perhaps unexpected prestige. Not only do they now possess increased value as coveted scientific specimens, with more minute detail than ever being exposed to scholarly observation—including, on occasion, soft-body parts such as antennae, gills, and legs—but they have also emerged as beautiful pieces of natural art that have attracted a surprisingly large and well to do mainstream audience. Getting a trilobite ready to attain such a lofty paleontological plateau, however, is far from easy. Each unprepared specimen presents its own unique set of problems and proclivities.

Some trilobites emerge from Canadian quarries where the rock is so hard that proper preparation—the kind performed without doing irreparable damage to the specimen's oft-fragile calcite exoskeleton—is nearly impossible. Others may hail from thinly bedded sedimentary outcrops in Ohio where the demarcation between trilobite shell and matrix is, at best, marginal. And still others may be found in brittle Estonian shale so easily shattered that it turns the initial stages of the preparation process into little more than assembling a sedimentary jigsaw puzzle. These challenges come with the Paleozoic territory, and the modern breed of trilobite preparators seems to have a handy solution for just about any problem that may be thrown their way.

"Working with trilobites is almost always a challenge," said Ben Cooper, recognized as one of the world's leading trilobite preparators. "Even the jobs that would initially appear relatively easy can suddenly turn difficult when you encounter something unexpected, which could range from a

crinoid stem laying directly on top of the trilobite to a pygidium that has become detached from the rest of the specimen. You need to keep a bag of 'tricks' handy—everything from glues to resins to epoxies that can solidify the matrix, fill in missing shell, and fix what you might accidentally break."

Those who spend their professional lives preparing trilobites for study, sale, or display are an admittedly unusual lot, with a significant segment of these men (and, yes, they are *all* men) determined to both walk and work to the beat of their own drummer. Many might best be described as introverted eccentrics, artistically inclined independent thinkers, and independent doers. Most seem to be nocturnal, preferring to ply their prepping craft long after those who follow a more traditional 9 to 5 work schedule have turned their backs on another day. It is in the relative quiet of the wee hours that these matrix-manicuring magicians do their best to transform even the most initially obscure, geologically obtuse trilobite specimen into a true Deep Time treasure. Sitting for prolonged sessions with their eyes glued to the lenses of their high-powered binocular microscopes, with their skilled hands often buried in the confines of well-ventilated prep cabinets (the fine-grained silicate powder used to remove surrounding rock from trilobites can cause major lung problems if inhaled over long periods of time), these guys seem happiest with only the ever-present whine of their air-abrasive machine to keep them company.

Despite their dexterity and dedication, dealing with such unusual men can, at times, be more than

NOVAKELLA COPEI FORTEY AND OWENS, 1987
Middle Ordovician; Lower Llanvirnian Series, Abereiddian Stage; Hope Shales; Near Minstrelsy, Shropshire, England, UK; 12.2 cm

This is perhaps the largest complete cyclopygid trilobite ever found. Unfortunately, the wraparound eye that distinguishes this genus is not present on this specimen.

RERASPIS PLAUTINI (SCHMIDT, 1881)
Upper Ordovician, Caradocian; Kukruse Regional Stage; Viivikonna Formation; St. Petersburg region, Russia; 2.1 cm

This is one of the few complete specimens known of this diminutive Eastern European species. Only in recent years have preparation techniques advanced enough that detailed features of these trilobites can be fully revealed.

a bit frustrating for those who collect and study trilobites. Prep guys rarely answer the phone (even when they're awake during daylight hours), usually don't respond to emails or texts, and motivating them to finish a long-overdue specimen can often turn into a true battle of wills . . . and patience. It has often been jocularly said of one talented midwestern prep master that if you send him a trilobite on Monday he'll get it back to you by Friday—just that the Friday in question may be two or three years hence. Another legendary prepper has become renowned for renaming the species (if not the genus) of every other specimen he works on—even if the taxonomic nomenclature for that particular trilo-type has been widely accepted for decades. Yet another has become infamous for literally putting out the "gone fishing" sign every time there is a salmon run near his northwestern home, even if he is in the midst of an on-deadline project for a major museum. Despite their occasionally curious and almost always unpredictable nature, without the creative work put forth by the two dozen or so individuals worldwide who've dedicated their careers to the art of fossil preparation, life simply wouldn't be the same for any of us who revel in those ancient arthropods called trilobites. By taking raw, fossil-bearing sedimentary stone, and adeptly turning its contents into objects of natural beauty and scientific importance, these prep maestros have found a unique way of *almost* bringing 500-million-year-old creatures back to life.

"Those of us who prepare trilobites generally love what we do," Cooper said. "You never know when you might reveal something that is new to science. Sometimes you find yourself cranking out a lot of common specimens, and you know you're doing it just to pay the bills. But sometimes you get the chance to work on something special. And when you're sitting there early in the morning, staring into a pair of trilobite eyes that you've just uncovered, there's no question that there's nothing else in the world you'd rather be doing."

EVERY TRILOBITE TELLS A STORY

There are those among us who can gaze on the fossilized remains of a trilobite and see little more than the vestige of a long-gone creature that has been strikingly preserved in sedimentary stone. Such a vision is certainly accurate from a purely aesthetic perspective; after all, trilobite fossils are both inherently appealing and endlessly fascinating, each one an invaluable relic drawn from

an incredibly distant time. But from a somewhat more intuitive point of view, that image of a trilobite embedded in its encasing rock cocoon serves as only the first chapter of a long and captivating story. If you look closely, every trilobite fossil tells a unique tale. Whether it's a 500-million-year-old *Hemirhodon amplipyge* bearing a potentially fatal bite mark or a 450-million-year-old *Primaspis crosotus* appearing forever frozen amid a fossilized field of crinoid stems, each example of these primeval life-forms possesses the capacity to provide a captured in time glimpse of a long-gone undersea kingdom. With a trilobite in hand, and a healthy imagination at play, each of us can partake in a private sojourn back to a time when trilobites filled the planet's Paleozoic oceans.

As modern preparation techniques have become increasingly efficient, more explicit details of trilobite anatomy have come to the fore. Corresponding theories concerning the lifestyles of these intriguing invertebrates have also emerged, with each discovery adding immeasurably to the stories every trilobite fossil can reveal. Morphological features such as 3-centimeter-long cephalic spines and wraparound compound eyes provide bold evidence of trilobite defensive postures and unprecedented evolutionary advances. Recently detected signs of fossilized eggs, gills, and gut tracts have afforded both scientists and collectors with previously unavailable, and unimagined, insights into the long-gone world inhabited by these primitive arthropods.

RIELASPIS ELEGANTULA (2 MIDDLE PRONES, 3 CM EACH) (BILLINGS, 1866); *NUCLEURUS SP.* (VENTRAL ON LEFT, 2 CM); *ENCRINURUS DEOMENOS* (ADJACENT TO LARGE CRINOID STEM) (TRIPP, 1962)
Silurian; Jupiter Formation; Eastern Canada

These trilobites hail from one of the most out of the way locales in North America. Such faunal associations reveal a dramatic story of trilobite lifestyles some 420 million years ago.

On occasion a carefully prepared trilobite may too closely resemble a piece of contemporary sculpture—especially those twenty-first-century specimens that sit pristinely perched on a perfectly manicured pedestal of half-billion-year-old matrix. Such extreme examples of the prep process run the risk of completely removing the trilobite in question from its paleontological context, subsequently obliterating much of that fossil's intrinsic story. Indeed, totally isolating a specimen from any accompanying fauna, flora, or sedimentary striation often makes that trilobite appear like a fish (or should we say an arthropod) out of water. Many of the recently prepared trilobite examples emerging from Morocco and Russia have been subjected to this brand of drastic display, being presented as beautiful yet strangely disconnected organisms—works of ancient art rather than scientifically significant specimens. But in stark contrast, some fossils—especially those showcasing multiple trilobites on a single slab or depicting these highly communal creatures interacting with starfish, cystoids, or crinoids within their undersea realm—present the scenario of prolific, active, social animals, each fully integrated into their timeless marine environment.

When taken collectively, each of the 25,000 trilobite species that arose during their 270-million-year swim through the Paleozoic possesses the capacity to reveal an amazing and unique story. Every trilobite fossil, from the most common *Elrathia* to the most exotic *Probolichas*, serves as a vital cog in the complex "machine" of evolution, and each tale they tell sheds new light on the darkest depths of Deep Time—the time when trilobites ruled the seas.

FAKE TRILOBITES

Apparently not all trilobites were created equal. Today, thanks to modern preparation techniques,

a select few of these ancient arthropods have emerged as pristine examples of Paleozoic preservation, half-a-billion-year-old fossils with flawless exoskeletons, ideal symmetry, and exquisitely detailed eyes. But for every such trilobite that may now proudly rest in a museum display or private collection, there seems to be a distinct doppelganger—a specimen that presents a far less noble and far more contemporary heritage. These manufactured monstrosities may look good on the surface; in fact, to the naked eye they may initially appear to exhibit nothing less than primordial perfection. Lurking just underneath their shiny black, brown, or tan outer coatings, however, are *not* the fossilized remains of creatures that dominated the world's oceans for more than 250 million years. What may at first seem to be an exotic example of the ancient trilobite line may actually be nothing more than the nimble-fingered concoction of a "factory" laborer operating deep within the heart of a faraway land.

Over the past 30-plus years, a thriving and apparently still expanding side industry has grown up around trilobites—one where native craftspeople, often working in rural villages without even the benefit of electricity, create their own brand of highly realistic fossils from mud, plastic, rubber, or just about any other reliably pliable compound upon which they can lay their artistically inclined hands. These unsavory practices have long been an accepted or at least acknowledged part of some trilobite transactions, especially those stemming from the paleontological stronghold of Morocco.

NELTNERIA JAQUETI NELTNER AND POCTEY, 1949

Lower Cambrian; Issafen Formation; Morocco; 15.2 cm

Species such as this were famously described (and illustrated) in the 1959 trilobite treatise. But only in the twenty-first century have complete specimens been discovered amid the hills and valleys of southwest Morocco.

MOROCCAN HANDIWORK

Here a scutellid cephalon has been combined with a lichid pygidium. The resulting chimera is little more than a confusing eyesore.

Since that nation's fossil trade first blossomed on the international scene in the late 1980s (featuring a diverse assortment of primal items, including crinoids, ammonites, dinosaur bones, and shark teeth, in addition to trilobites), North African artisans have become notorious for utilizing everything from local mud and mortar to automobile repair putty (bondo) to carefully construct a veritable sunbaked smorgasbord of faux trilobites. These "fossils" range from what may initially appear to be large Middle Cambrian *Paradoxides* and *Cambropallas* specimens to a mind-numbing variety of delicately barbed Devonian trilo-types. Often the work of these fabricators is so compelling—with their original molds crafted from top-grade authentic examples—that

RUSSIAN TOMFOOLERY

Some St. Petersburg artisan apparently found a disassociated lichid head and cheirurid body and decided to somewhat haphazardly combine them.

even university-trained trilobite experts have routinely been fooled.

"There's no question that discerning fake trilobites from real ones has become something of a problem," said David Rudkin, formerly of Toronto's Royal Ontario Museum. "The Moroccans have become amazingly skilled at what they do, and I've seen some very smart people be taken in by their efforts."

As might be expected, when directly confronted about these egregious fossil-generating activities, most involved with the Moroccan trilobite trade quickly plead innocence, or at least ignorance. At times, it seems as if virtually all Moroccan fossil dealers share some clandestine, omerta-like code of silence when it comes to discussing their

nation's more overt paleontological misdeeds. Yet even those merchants willing to address this ticklish issue often can't help but further cloud the deception. Rather than acknowledging at least some degree of wrongdoing, they do their best to dismiss it, explaining that rather than totally creating fossils from scratch their efforts are designed to merely enhance the appearance, and marketability, of certain rare trilobite specimens.

When further pressed on the matter, some dealers will state that their realistic copies are devised expressly to accommodate the indiscriminate tastes of the tourist crowds that prowl Casablanca and Marrakech. Others use every trick in their well-rehearsed verbal repertoire to avoid admitting that such handiwork is done to deceive the supposedly knowledgeable scientists and collectors who travel for long hours, over difficult terrain, to reach their desert-hugging outposts. The evidence, however, says something quite different. Not only are these merchants painfully aware of what they are doing, but the substantial profits generated by their misleading efforts do little to deter them from continuing, and perhaps even escalating, such activities in the future.

Why do so many Moroccan merchants exert so much energy fabricating trilobites from available, often unnatural resources when they're sitting atop an apparently unlimited supply of genuine fossil material right there in their homeland? A recent visit to the trilobite quarries that ring the Sahara border town of Alnif revealed a multitude of workers digging up authentic Devonian, Ordovician, and Cambrian trilobites, with many openly offering to sell these unprepared specimens on the spot for extremely reasonable prices. It appears that the answer as to why many Moroccan fossil merchants don't routinely avail themselves of these handy trilobite reserves is rather simple, and somewhat sensible. After all, in a remote part of a nation where cars are still uncommon, it is far

**SINOSAUKIA DISTINCTA
(ZHOU, 1977)**

Upper Cambrian, Furongian; Jingxi
Section, Sandu Formation; Guole,
Jingxi County, Southwestern Guangxi
Province, China; 13.1 cm

This colorful specimen represents one
of the largest complete examples of
this trilobite yet found.

easier to sit at home and manufacture a marketable product by the
dozen than to venture into the field and risk both time and money trying to uncover, or purchase, trilobite specimens that may, after hours of
careful preparation, turn out to be less than perfect.

Despite the seemingly unstoppable tide of fabricated fossil material
still emanating from North Africa, the battle of "real vs. fake" is apparently not yet lost. It is likely that there are currently more blatantly
fraudulent Moroccan trilobites than ever on the world scene—as any
visit to eBay will quickly reveal—but it now appears as if the Paleozoic
tide may slowly be turning in favor of the trilobite enthusiast. In recent
years, a growing number of higher-end Moroccan fossil suppliers have
begun importing expensive preparation equipment—along with the

generators needed to run them—to aid in their increasingly successful attempts to take the marketing of legitimate North African trilobites to the next level of commercial and academic acclaim. Indeed, their work with genuine specimens—including scarce spinose species such as *Kohliapeltis rabatensis* and *Comura bultyncki*—has now reached such a state of artistry that it has served to make these top-of-the-line examples among the most coveted trilobites on the entire planet.

"It's almost become expected that when you see Moroccan trilobites at major fossil shows around the world—let alone little rock shops in your local town—that at least some of them are going to be fake," said Bill Barker, a leading importer of Moroccan fossils. "But I must add, in recent years some of the biggest Moroccan merchants have turned a major corner. They've realized that quality will make them more money than fakery, and they've begun producing some simply amazing, and totally real trilobite specimens, many with freestanding spines. But sadly, too many Moroccan dealers are still glued at the hip to their bondo supplies."

Even with the significant recent progress made in lending a more honorable aura to the Moroccan trilobite trade, it is still impossible to attend any of the major fossil shows held around the globe and not be continually confronted by North African dealers selling nothing but artfully assembled imitation arthropods. Some of these merchants are cunning enough to disguise their efforts by rather grandly displaying a few select counterfeit specimens in well-lit, glass-enclosed cases, at least attempting to create an ambiance of quality and exclusivity. Others are pure businessmen, presenting tables filled with row after row of identical *Dicranurus*, *Ceratarges*, or *Psychopyge* trilobite casts with nary an effort made to hide their dastardly deeds.

Despite the distaste that all serious collectors hold for such deceptive practices, many of these dealers in fake fossils present their goods at reasonable prices, often as low as $20 per piece. Thus, if one is willing to look at these "bondo bugs" as nothing more than craftily constructed curios, or scientifically accurate replicas, they may be viewed as serving the somewhat beneficial purpose of promoting an available and affordable avenue into the paleontological world—especially for kids and novices. When the line in the Sahara sand is crossed, however, and patently bogus Moroccan material is offered at top-dollar prices as the Real Thing, an unwritten rule—as well as a variety of international laws—has been broken, often with nasty results.

Reports of fistfights and heated verbal jousts (in a variety of languages) have become part of fossil show lore. Many unwary customers have turned understandably angry when later informed of the artifice foisted upon them—especially after some believed they had just acquired an incredibly rare trilobite species for an unbelievably reasonable price! Once they discover that *they* were the ones who had been duped—rather than their supposed Third World pigeon—they invariably return to confront their Moroccan salesman, only to be informed, with minimal fanfare, that all deals are "final." Needless to say, local authorities—as well as undertrained, understaffed show security guards—have been continually kept on their toes by such actions and reactions.

"There are still people who come to these shows who don't realize that many of the Moroccan trilobites they see are questionable," Barker said. "Of course, there are now a lot fewer people who fall into that category than 20 years ago. Then it was still something of a scandal, and more than a few tempers were lost as deals went awry."

As many trilobite collectors have sadly come to learn, Morocco is far from the only place on Earth where such paleontological tomfoolery is taking place. In recent years, a similar but somewhat more sophisticated method of fossil fabrication

**POSTIKAOLISHANIA
JINGXIENSIS ZHU, 2005**

Upper Cambrian; Sandu Formation;
Guangxi Province, China; 4.3 cm

This intriguing specimen reflects the
diverse trilobite fauna currently being
uncovered in southern China.

has been established by those diggers, preppers, and merchants who operate in and around St. Petersburg, Russia. Local workers, many of whom also actively search the neighboring Ordovician quarries for authentic fossil material, have begun producing a line of trilobites that have been augmented—if not totally constructed—through the use of high-tech polymer plastics and resins. In the hands of inventive and often knowledgeable Russian artisans, these space-age ingredients are made to perfectly match in color and texture the lovely, caramel-hued carapaces presented by the impressive line of local trilobites. Not by happenstance, many of these creatively enhanced trilobites represent the exact same exotic species that sit at the top of many collectors' "want" lists. Unwary enthusiasts too often find themselves spending

CAMBROPALLAS AFF. TELESTO

Middle Cambrian; Jbel Wawrmast Formation; Morocco; 6.1 cm

This is a beautifully preserved—and real—example of one of the most often faked trilobites emerging from North Africa.

Asery Horizon formations. In fact, examples of a surprising number of exceptionally rare (and anatomically accurate) Russian cheirurid and lichid species exist in complete form only as these carefully manufactured arthropod amalgams. And to add even further frustration, many of these fossiliferous fusions are virtually impossible to detect—even with the help of high-power magnifiers and black lights. Recognizing the differences between one of these reconstructed trilobite specimens and a pristine "original" can, at best, be difficult, and collectors should note that most trustworthy Russian dealers (if that itself is not something of an oxymoron) will usually present their premier frankensteined creations at a somewhat lesser price than the genuine Paleozoic article.

"The Russian dealers are very savvy," said Sam Stubbs, a Houston-based trilobite collector. "If a deal they're presenting to you on a certain trilobite appears 'too good,' then you must assume that it is."

At the opposite end of the ambitious fossil fabrication spectrum that in recent years has been employed so effectively by Russian trilobite merchants are the rather cumbersome efforts put forth by local workers in Bolivia and China. In quiet corners of these remote lands, crude yet often strangely captivating pieces of what might best be termed trilo-art have been manufactured and marketed—often by those who appear to possess little more than a fundamental knowledge of basic trilobite morphology. The vast preponderance of these flagrantly fake fossil specimens are relegated to small, tourist-oriented shops located in the outskirts of La Paz or Shanghai. Other examples, however, have made it all the way to the instant international marketplace provided by the internet, where poorly executed trilobite details can be easily obscured by fuzzy photo quality or intentionally bad lighting. To add yet another element to this oft-confusing scenario, when Chinese and Bolivian merchants make their increasingly

exorbitant sums on these palpably plastic examples—prices often humorously justified by the dealer as reflecting the cost of their time fabricating these deceptively faux fossils.

If such spurious practices weren't confusing enough for hobbyists struggling to keep up with the seemingly never-ending advancements in fossil fakery, the Russians have also begun specializing in another blatantly unethical, although perhaps slightly more acceptable, practice. Some St. Petersburg–based merchants now focus on compositing complete trilobites from the voluminous numbers of authentic disarticulated cephala and pygidia found throughout the area's 450-million-year-old infrequent appearances at major fossil shows in the United States or Europe, authentic trilobite specimens are often haphazardly mixed amid a flurry of instantly detectible frauds.

To their marginal credit, at least some of the Bolivian merchants exhibit a rather whimsical sense of humor through their fossil fabricating craft. They occasionally supply their two-part trilobite casts (which draw inspiration from their nation's renowned Devonian concretion-encased specimens, many of which continue to be the subject of major scientific study) with tiny carvings of human faces, arms, or legs. A recent purchase made by a visitor to a La Paz gift shop included a mud-cast trilobite replica complete with a tiny Christian cross adorning the center of its cephalon.

"Many of these merchants aren't that familiar with fossils," said Martin Shugar, a field associate with New York's American Museum of Natural History. "To most of them, trilobites are nothing more than another indigenous trinket, something to be marketed to the first person who shows the slightest interest."

In the wake of this international trilobite "conspiracy" where cleverly disguised fakery is still a concern (with some sources now using cutting-edge 3D printer technology), what are fossil enthusiasts to do? Should they sacrifice the enjoyment and enlightenment their hobby provides because of the understandable fear of being taken to the proverbial cleaners by unscrupulous—or oblivious—merchants from distant lands? Whether they're willing to admit it or not, everyone who seriously collects trilobites owns at least one specimen—and probably *many* more—that has been doctored to a greater degree than originally imagined. Whether they're created in a back room in Morocco or a basement in Bolivia, these fraudulent fossils are now more pervasive than ever. Even in this savvy, sophisticated, computer-driven age, most museums unknowingly possess and showcase out and out phony Paleozoic relics. Indeed, a few years back Toronto's prestigious Royal Ontario Museum presented an exhibit called *Fakes & Forgeries* that detailed this exact scenario, and included an alarming number of trilobites in their display.

Perhaps such uncertainty is just the price both the collecting and scientific communities must pay for doing business in today's frenzied fossil market. As costs for items such as rare trilobites continue to soar, and new fossiliferous locations open in countries both near and far, the chances of circumspect material hitting the world stage are now stronger than ever. Rather than throwing up their hands in disgust and walking away, trilobite hobbyists should be prepared to enter these upcoming "battles" armed with both a bit of discerning insight and a well-honed degree of skepticism. A good hand lens and a small UV black light (which may help detect reconstructed or fabricated anatomical features) can assist you in differentiating a fraud from a fantastic specimen. After all, air bubbles are hard to hide . . . even with industrial grade bondo. To avoid potential problems, it's also essential for any interested party to possess a healthy dose of basic street smarts, a quality that should allow a collector to look both a trilobite and a merchant squarely in the eye and sense who is, in fact, getting the best of whom.

TRILOBITE COLOR PATTERNS

Whatever your level of interest in the trilobite world, one thing should quickly become apparent—the fossilized forms of these ancient

PARVILICHAS MAROCHII CORBACHO AND VELA, 2013
Lower Ordovician, Floian; Upper Fezouata Formation; Drâa Valley, near Zagora, Morocco; 6.4 cm

The dramatic coloration displayed on this rare specimen is due to the abundance of mineral-rich sediments at this site.

arthropods display an amazing diversity of color. Depending on where they were found and the minerals that have most influenced their process of preservation, the rock-hard carapaces of these Paleozoic relics can appear in tones of black, white, tan, green, or brown, along with an occasional tinge of red, orange, or yellow. Some specimens even have a mottled shell pattern akin to a calico print.

From the golden hues that distinguish the myriad species unearthed in the Ordovician layers surrounding Russia's Volkhov River, to the ghostly alabaster preservation that defines much of Portugal's Valongo Formation, to the charcoal gray that characterizes the Silurian material drawn from upstate New York's Rochester Shale, virtually every color contained in our planet's natural palette can be seen in the calcified remains of trilobite exoskeletons. But a key question remains: Do any of these shades represent the colors these incredible invertebrates may have displayed during their long-ago lifetime in the primeval seas?

Much like modern birds, butterflies, or tropical fish, it's not difficult to imagine trilobites of different genera traversing those antediluvian oceans exhibiting a wide variety of contrasting tones. Indeed, distinctive coloration appears to be a recurrent characteristic in the animal kingdom—and what is true today may have been true when trilobites crawled through the seas some half a billion years ago. Perhaps those Paleozoic color patterns were part of an intricate mating ritual or were used as a defensive mechanism or as a means of finding other brightly tinged species members hidden amid the murky marine depths. It is unlikely that we will ever uncover a definitive answer to this compelling question about these early life-forms, but perhaps we will learn more. A smattering of recent evidence has begun to shed a few dim yet promising rays of light on the

ELDREDGEOPS RANA (GREEN, 1832)
Middle Devonian (Givetian); Hamilton Group; New York, United States; 5.7 cm

It has been speculated that the small dots that adorn this trilobite's carapace may represent a real-life camouflage pattern.

Photo courtesy of Markus Martin

somewhat controversial subject of trilobite color patterns and whether they actually—and accurately—exist within the fossil record.

We know so little about the color ornamentation of trilobite exoskeletons for a very good reason. Hundreds of millions of years after their demise, the various calcite-infused pigments that created those colors in life have been replaced by the invading minerals that are essential to the fossilization process. But a small number of recently studied specimens, including examples of *Scabriscutellum scabrum* and *Thysanopeltis acanthopeltis* recovered from the Devonian-age Ahrdorf Formation of Germany, display what appear to be well-defined striated marks adorning their disarticulated pygidia. These bold designs have persuaded some scientists to suspect that they may reflect true-life trilobite color patterns, although other paleontologists remain reticent to embrace such a radical concept, believing that these unusual shell markings may be a direct result of the postmortem mineralization process itself.

The planet's sedimentary soils have also possibly preserved another in-life characteristic of the trilobite carapace—spots. What evolutionary role these small, dark, circular features may have played as trilobites darted in and out of sun-splotched coral reefs, or languished partially buried along the seafloor, is still very much open to speculation and debate. However, the presence of these distinctive markings now appears irrefutable. Examples of the abundant phacopid *Eldredgeops rana* found within the Devonian deposits of western New York State exhibit a series of distinct, diminutive black dots on their fossilized outer shells. It has been theorized by some in the academic community that these spots represent a type of primitive camouflage, although others feel that they may simply be the dorsal reflection of the trilobite's internal muscle attachments.

Many of the trilobite fossils adorning private collections and museum shelves display a veritable rainbow's range of colors, but none of those shades is likely to represent the trilobite's actual, in-life hue. Perhaps one day we will uncover a Paleozoic layer in which the trilobite material reflects a natural shell coloration. Until that time, all we can do is speculate and wonder about this intriguing mystery of the ancient past.

TRILOBITES IN HISTORY

In the spring of 1886, a group of intrepid archaeologists began exploring a series of limestone caves near Arcy-sur-Cure, a small French community about 200 kilometers southeast of Paris. They were in search of human relics, including spear points and bone utensils dating back to the Pleistocene Epoch. What they found was something considerably older—and totally unexpected. Inside one of the caves, in a layer that later dated back some 15,000 years, they recovered a 450-million-year-old Ordovician-age trilobite. That discovery alone may have been noteworthy, but it was made even more significant when it was determined that the fossil featured a hand-drilled hole through its pygidium, a detail that allowed the specimen to be worn as an amulet or fetish. From its smooth but rather weathered appearance, it was immediately evident to these explorers that this trilobite had been a treasured totem for the ancient tribe that once inhabited the cave, a locale that subsequently came to be known as the *Grotte du Trilobite.*

Trilobites have long played a role (albeit a relatively minor one) in human history. Despite the roughly 250 million years which separate the demise of that arthropod line and the rise of our own species, there has been a surprising degree of interaction between trilobites and *Homo sapiens* throughout our disparate spans on Earth. Some of our ice age ancestors in Europe apparently revered trilobites; so did a variety of Native American tribes located in the southwestern desert. Members of the Ute tribe routinely carried small, 500-million-year-old *Elrathia kingii* specimens in their medicine pouches both to provide protection from enemies and to ward off evil spirits. And petroglyphs that seemingly depict an indeterminate species of trilobite have been found adorning cliff walls in southern

Utah. Some archaeologists believe these fanciful images could be hundreds, if not thousands, of years old. Evidence of this tribal fascination with trilobites extends from British Columbia all the way down to Australia, where talisman featuring trilobite parts of varying sizes and shapes have been discovered in various ancient Aboriginal sites dotting the southern half of that island continent.

In what some might label more "advanced" societies, vague descriptions of strange "stone scorpions" appear in European writings dating back to the third century. At roughly the same time, references to "swallow stones" (featuring the disarticulated pygidia of the Cambrian trilobite *Neodrepanura premesnili*) can be found in certain Chinese documents. And it is known that trilobites were often treasured throughout eastern Asia a thousand years ago, appearing as decorative items adorning places of honor in the most cultured homes.

Despite these early manifestations of trilo-centric interest, the true study of these primal invertebrates didn't begin in earnest until the last years of the seventeenth century. It was then, in England, that Reverend Edward Lhuyd made the first direct mention of a trilobite in scientific literature. His description was of something he called a "flat fish," which centuries later would be identified as the trilobite *Ogygiocarella debuchii*. In the academic journal *Philosophical Transactions of the Royal Society*, Lhuyd also presented carefully

OGYGIOCARELLA DEBUCHII (BRONGNIART, 1822)

Ordovician, Uppermost Llanvirn Series; Middletown Formation, Meadowtown Beds; Betton Wood Quarry, Shropshire, England, UK; 8.5 cm

During the latter years of the seventeenth century, this was the first trilobite described in scientific literature. At that time, it was labeled a "flat fish."

NEODREPANURA PREMESNILI (BERGERON, 1899)

Middle Cambrian; Kushan Formation; Shandong Province, China; 3.2 cm

The distinctive (and virtually always disarticulated) pygidia of this species were described as "swallow tails" in ancient Chinese writings. The first documented complete example of this trilobite was uncovered in 2006.

constructed drawings of his find, which represent the first widely dispersed images of a complete trilobite.

Half a century later, in 1750, Englishman Charles Lyttleton conducted the initial research on a specific trilobite species, submitting a paper to the Royal Society of London on the famed Dudley Locust, now known as the Silurian trilobite *Calymene blumenbachii*. His description of this "petrified insect" fueled a firestorm of controversy across the nation and, for all intents and purposes, ignited the era's scientific debate regarding the role trilobites played within the animal kingdom. However, nearly another century would go by before the renowned Scottish geologist Roderick Murchison laid the groundwork for future trilobite research with his historic volume, *The Silurian System.* That manuscript, released in 1839, intricately described the fossil fauna found throughout Britain, and its subsequent notoriety turned Murchison into a sensation in Europe's most erudite circles.

Murchison's celebrated efforts led directly to the ensuing, pioneering, and oft-lauded trilobite-related research conducted by the likes of the French naturalist Joachim Barrande and the American adventurer Charles Walcott. Their accomplishments, along with the subsequent endeavors of many others, including contemporary paleontologists such as Niles Eldredge and Richard Fortey, have added important chapters to the evolving history of the world's favorite fossilized arthropod, the trilobite.

THE STRANGEST TRILOBITES

Trilobites survived for more than a quarter of a billion years—plenty of time for them to develop into the 25,000 species recognized by science. But amid this mind-boggling degree of biological diversity are some species that are so unusual, so strange, so brazenly bizarre that they merit special morphological mention. These are creatures so alien in appearance that they could easily serve as featured attractions in any upcoming Hollywood sci-fi spectacular—but few observers would believe that such out of this world lifeforms could possibly have existed. But they did, and they were among the first rulers of the Earth's Paleozoic seas.

Imagine the likes of *Actinopeltis globosus*, a Devonian trilobite with a perfectly symmetrical "ball" perched atop its glabella. Picture

other Devonian trilo-types such as *Walliserops trifurcatus*, with a prominent, trident-like fork extending from the front of its head, or the aptly named *Dicranurus monstrosus*, with two menacing "horns" sweeping back from the top of its cephalon. Or how about such Ordovician species as *Asaphus kowalewski*, with eyes sitting atop 5-centimeter-long stalks; or *Cyclopyge bohemica*, which possessed a single, multilensed eye that literally wrapped around the entire front of its head; or *Probolichas kristiae*, a trilobite with a sword-like cephalic "probe" that preceded it through its oft-murky marine world.

The heterogeneity of trilobite body configurations is almost impossible to fathom—even for those who have long collected and studied these primeval marine inhabitants. Amid those assorted anatomical alignments are some surrealistic features that even the likes of Salvador Dali would have had difficulty envisioning: *Erbenochile issimourensis* had a shading "brow" atop its thick, 2-centimeter-high eye stacks; *Odontocephalus ageria* sported a frilled "cowcatcher"—possibly used to stir up and then filter seafloor sediments—emanating from the front of its cephalon; and *Asaphellus cuervoea* had genal spines so long and wide that they may have functioned like wings, allowing this species to glide through the warm bays and basins it called home.

Scientists all agree that each of these unusual morphological features benefited their host

PROBOLICHAS KRISTIAE CARLUCCI, WESTROP, AND AMATI, 2010
Upper Ordovician; Bromide Formation, Pooleville Member; Arbuckle Mountains, Oklahoma, United States; 9.3 cm

From every imaginable perspective, this is one of the rarest and strangest trilobites in the world. Paleontologists are just now beginning to ponder the possible uses for its distinctive cephalic "probe."

trilobite in some significant manner during the animal's daily quest for survival. It has been speculated, for example, that the "ball" atop the head of *Actinopeltis* may have been used as a flotation aid while the trilobite navigated rough seas; others surmise that the same anatomical feature might well have functioned as an egg sack. The intimidating trident attached to *Walliserops* may have served a variety of purposes: an anchor during storms, a grasping device while mating, a defense mechanism, or even a tool used to filter food.

It is clear that trilobites were highly adaptable animals, organisms capable of evolving rapidly and effectively to maximize any ecological advantage. Although tridents, horns, eye stalks, and nose sacks may seem highly unusual to many of us, development of such outlandish characteristics allowed the trilobite class to leave behind a fossil legacy perhaps unequaled by any other life-form in our planet's protracted history.

TIME AND TRILOBITES

The late, great science fiction writer Isaac Asimov is often credited with recognizing the human brain's somewhat limited capacity for comprehending large numbers. In his award-winning discourses, the eloquent New Yorker would frequently contend that each of the world's cultures—whether based in the wilds of Borneo or on the mean streets of Boston—display distinct words in their dialect representing the numbers one through five. He would then go on to explain that a select few of those cultures also possess a single term within their collective vocabulary designed to represent any greater sum. It is a basic and perhaps highly utilitarian concept—"many" in lieu of unnecessary specifics. Of course, the inherent flaw in this line of reasoning is obvious—there could be six enemy warriors coming over that hill to attack your village

or six hundred. In certain circles, communicating such vital information in a timely and efficient manner could prove quite problematic.

The fact is that prodigious numbers, or at least those higher than five, have always had a way of boggling the minds of our still rather primitive species. Oh, sure, we can all pretty much grasp the significance of a $20 lunch bill, or follow the point-by-point progress of an NBA game. We are certainly able to easily pronounce figures such as a million, or even a *trillion*, perhaps when discussing a lucky friend's Wall Street bonus or the gross domestic product of Germany. But for whatever reason, the true enormity of such numbers inevitably fails to fully click within the depths of our still-evolving craniums. Perhaps our brains simply weren't wired to ponder the number of stars in the sky, the multitude of sand grains on a beach, or the Deep Time age of the universe. But such a restrictive notion certainly hasn't stopped some of us from trying to do exactly that.

Those of us who collect and study trilobites have never been a group particularly intimidated by the concept of large numbers. After all, in a field where tens of thousands of species are scientifically recognized, and hundreds of millions of years represent the standard means of time measurement, a well-honed "feel" for mind-numbing numbers appears to go right along with the Paleozoic territory. Coming in daily contact with primitive trilobite species such as *Fallotaspis typica*, *Kootenia spencei*, and *Bristolia insolens*, whose

DIKELOCEPHALUS MINNESOTENSIS (OWEN, 1852)

Upper Cambrian; St. Lawrence Formation; Sauk County, Wisconsin, United States; 12.3 cm

Until its dismantling in 2012, this specimen long served as the centerpiece of the Smithsonian's famed trilobite display. The trilobite was originally "found" serving as a doorstop in a midwestern home in the 1950s.

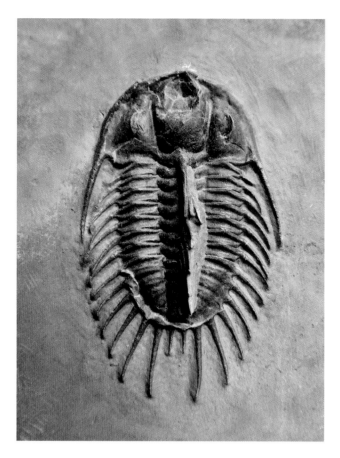

***KOOTENIA SPENCEI* (RESSER, 1939)**
Middle Cambrian; Langston Formation, Spence Shale Member;
Antimony Canyon, Utah, United States; 5.3 cm

Kootenia rank among the most often collected trilobites found within the rugged Spence outcrops, although complete specimens are scarce.

in the fabric of time mean little. In fact, for many of the more paleontologically inclined among us, a phrase such as "give or take a million years" is an accepted and welcomed part of everyday lexicon. There seems to be little doubt that those who choose to sequester themselves in a realm where trilobites dominate their thoughts, deeds, and actions are willing to stare the great abyss of time squarely in its fanciful face, and live to tell about it.

"Even if you've been interested in fossils all your life, it's impossible to hold something like a trilobite in your hands and not be somewhat astounded," said Martin Shugar, a noted New Jersey-based collector. "Invariably, if you have any imagination at all, you eventually get caught up with wondering what was happening on this planet back when the creature you're holding was alive."

The earliest members of the trilobite lineage date back some 521 million years, and those pioneering invertebrates subsequently existed in the seas for the next 270 million years. Yet the truth is that the unimaginably long tenure enjoyed by these primeval arthropods actually covers only a small portion of Earth's history. With all our cutting-edge technology now confirming a 4.54-billion-year-age for our world, even a rudimentary understanding of mathematics leads us to the conclusion that, despite their protracted passage through the Paleozoic, trilobites spent less than 5 percent of the planet's total history amid the Earth's ever-changing evolutionary chain.

It's somewhat surprising that our acceptance—let alone the comprehension—of such imposing numbers is a relatively recent phenomenon. Just

half-billion-year-old origins date back to the very beginnings of complex life on our planet, seems to place trilobite-centric folks squarely within their Cambrian comfort zone.

Whether it's due to the atypical nature of their peculiar interest, or is merely the by-product of a healthy imagination, those fascinated by trilobites appear to possess a decidedly different lens through which to view the movement of time. For most of the globe's residents, the passage of a year, let alone a decade, is cause for remembrance and celebration; but for certain card-carrying members of the fossil fraternity, such minimal disturbances

***HUNGIOIDES MIRUS* LU, 1975**
Ordovician; Arenig Seriesl; Dawan Formation; Hubei Province, China; 7 cm

This rarely seen species is strikingly similar to one emerging from the Valongo layers of Portugal, now some 10,000 kilometers away.

a few centuries ago, many within society's most erudite circles blindly accepted the "fact" that the Earth was all of 6,000 years old. Some of the era's leading pundits even postulated that an exact date (October 23, 4004 BCE) could be established upon which this Third Stone from the Sun first began its dizzying trek through the cosmos. Few could, or would, dare debate the scientific merits behind such a claim. Their understandable fear was that the wrath of prominent political groups and religious orders would be the most likely outcome of any such incendiary confrontation. Indeed, throughout the British Isles during the late 1700s, a death penalty could be imposed upon anyone unwise enough to openly debate accepted religious doctrine, almost certainly including the subject of the Earth's antiquity. Anyone who encountered a trilobite during these scientific dark ages viewed it merely as an inexplicable curio or perhaps a supernaturally tinged "insect made of stone." After all, with the world only a few thousand years old, what true significance could these diminutive, rock-hard relics possess?

As recently as the early years of the twentieth century, many biblical scholars and a surprising number of tenured academics still readily accepted such an absurdly modern age for our planet's birth. Despite the best efforts of scientists ranging from Darwin and Murchison to Barrande and Walcott, a wide swath of mainstream Western society, whether in Europe or North America, still held on with a steadfast intensity to their misguided beliefs regarding a 6,000-year-old Earth. Even today, notwithstanding a stockpile of irrefutable paleontological and evolutionary evidence, a shockingly large number of vociferous souls still cling to these outdated (yet undeniably more easily digestible) concepts concerning the planet's age . . . or lack thereof.

These days we may choose to look askance at anyone willing to blindly accept such a seemingly ridiculous notion. However, those among us who have ever attempted to seriously ponder the 521 million years that have transpired since the emergence of trilobites, let alone the Earth's 4.54-billion-year history or the 13.82-billion-year age of the universe, are invariably and properly intimidated by such an impossible to comprehend task. There is unquestionably something awe-inspiring, perhaps even somewhat divine, cloaked within those expansive numbers, something that motivates us all to quietly reflect upon our own trifling place within the cosmic order.

For a species that has existed in its present form for perhaps 300,000 years, and whose entire direct ancestry can be traced back only 14 million years, we *Homo sapiens* can be an incredibly arrogant bunch. Sometimes we think we've got all these Cambrian Explosion, plate tectonic, and punctuated equilibria things pretty much figured out, with all the adjacent Snowball Earth and Big Bang mumbo-jumbo thrown in for good measure. But can we ever hope to fully digest the concept—let alone the content—of nearly five billion years of Earth history; or the quarter-billion years that have passed since trilobites last roamed ancient seafloors; or the 65 million years since dinosaurs walked the face of our world? To say the least, to do so would be an imposing task for anyone. But perhaps it's especially challenging for those who have chosen to make such cerebral concepts an essential part of their fossil-collecting hobby, or even the focus of their life's work.

Despite the scientific import that trilobites have long held, and the growing cultural gravitas that they currently enjoy as much-coveted collectibles, a conspicuous number of us continue to dismiss these ancient arthropods as little more than an inconsequential and failed evolutionary experiment—one that culminated in the seemingly inevitable fate of extinction at the end of the Permian, some 250 million years ago. Many others,

***EOHARPES CRISTATUS* ROMANO, 1975;
EODALMANITINA DESTOMBESI HENRY, 1965**

Ordovician; Llandeilo Series; Valongo Formation; Portugal;
Eoharpes cristatus: 3.1 cm; *Eodalmanitina destombesi*: 7 cm

This engaging "double" features two unusual Valongo Formation species.

face of Earth but recognize their fossilized remains as tangible proof of their role as one of our world's first, and most adept time travelers.

"There are many among both paleontologists and those that enjoy trilobites as a hobby who seem to have a fascination with the concept of time and its impact on evolution," said Shugar. "The late Stephen J. Gould was one of those. He spoke and wrote so eloquently on subjects that often touched on those exact topics. Many others have also written with passion and insight on similar issues, both in a fictional and nonfiction capacity."

Our understanding of time lingers among the greatest challenges still confronting the human mind. After all, time places elusive yet tangible limitations on our lives and ambitions while providing a murky depth to both our perceptions and realities. These days we can each expect to enjoy a 75-year life span, which compares with a 3-year life expectancy for a mouse, 30 years for a chimpanzee, 60 years for an elephant, and over 100 years for a giant tortoise. When contrasted with the 13.82-billion-year age of the universe, the 4.54-billion-year history of our planet, or the 270-million-year reign of trilobites, such numbers appear so small that they border on the insignificant. And perhaps they are. With a brain that too often forgets where we left our car keys or has trouble balancing a checkbook, maybe we humans were simply never *supposed* to consider weighty matters such as the longevity of trilobites. Just maybe, when everything is said and done, we will all find a degree of reassuring truth in what the old axiom (and Albert Einstein) has long told us . . . time is all relative.

however, note the impressive duration of their passage through geological history, as well as their incredible diversity, which saw 25,000 scientifically recognized species inhabiting virtually every available niche within their aquatic domain. These folks not only acknowledge the esteemed trilobite as one of the most successful creatures ever to roam the

FINAL THOUGHT

Trilobites emerged in the Earth's seas 521 million years ago. To provide some Paleozoic perspective on that impressive number, consider that counting from zero to 521 million, at a rate of one number per second, would take more than 16 years.

FALLOTASPIS BONDONI (NELTNER & POCTEY, 1950)

Lower Cambrian, Series 2, Age 3; Issendalenian Regional Stage; Zagora region, Morocco; 6.5 cm

This is a large example of one of the earliest trilobite species to be discovered within the fossil-rich soils of North Africa.

ACKNOWLEDGMENTS
Special Thanks

DURING MY THREE DECADES of collecting trilobites, certain people—whether it was their initial intention or not—have continually stepped forward to inspire my peculiar Paleozoic passion. These individuals include such fellow trilobite enthusiasts as Sam Stubbs, Ray Meyer, Bill Barker, Warren Getler, Tom Lindgren, Val Gunther, Perry and Maria Damiani, Phil Isotalo, Doug DeRosear, Jack Shirley, Robert Schacht, Richie Kurkewicz, Carles Coll, Greg Heimlich, Matt Phillips, James Cook, Carlo Kier, and Brian Whiteley. At one time or another, I have happily traded specimens, garnered knowledge, swapped gossip, and shared insight with each of these arthropod-obsessed associates.

Then there are those invaluable members of the fossil fraternity whose life's mission seems to be focused on uncovering trilobites in some of the planet's most inaccessible localities . . . and subsequently funneling many of these prize discoveries in my direction, often at *outrageous* prices! This list of intrepid explorers and extractors includes Jason Cooper, Dan Cooper, Markus Martin, Tom Johnson, George Ast, and Jake Skabelund, quite a few of whom are referenced—and even quoted—within these pages.

None of the specimens featured so prominently in *Travels with Trilobites* would appear anywhere near as visually appealing without the efforts of talented prep masters like Ben Cooper, Alf Cawthorn, Dave Comfort, Kevin Brett, Scott Vergiels, and Zarko Ljuboja, each of whom

KOOTENIA RANDOLPHI (ROBISON AND BABCOCK, 2011)
Middle Cambrian; Pierson Cove Formation; Drum Mountains, Utah, United States; 5.6 cm

This is a beautiful specimen of one of Utah's most collectible trilobites. Known for decades, it was renamed following a major scientific revision in 2011.

PROTOLLOYDOLITHUS NEINTIANUS WHITTARD, 1956 AND CNEMIDOPYGE PENTIRVINENSE KENNEDY, 1989

Middle Ordovician; Llanvirn Series, Shelve Formation, Stapeley Volcanic Member; Leigh, Shropshire, England, UK; (*left*) 5.7 cm, (*right*) 3.4 cm

Here is an unusual pairing of two relatively common members of the Shropshire trilobite fauna.

manages the seemingly impossible task of transforming slabs of raw, trilobite-bearing sedimentary stone into museum-quality showpieces.

Not only are all these fine *Homo sapiens* incredibly talented in their chosen fossiliferous field, they must each live with the burden of knowing that I consider them friends as well as colleagues. I send out a heartfelt thank you to every one of them.

And last, but certainly not least, special thanks go in the direction of Martin Shugar, my coeditor on the AMNH trilobite website, a notable collector in his own right, whose tireless moral support and bulldog-like determination played a major role in helping to turn this book from a half-baked notion into a fully baked reality.

(OPPOSITE PAGE) **FLEXICALYMENE SENARIA (CONRAD, 1841) AND CERAURUS PLEUREXANTHEMUS (GREEN, 1832)**

Upper Ordovician; Bobcaygeon Formation; Belleville, Ontario, Canada; matrix: 38 x 30 cm

Here is a massive plate covered by complete examples of two of eastern Canada's most prominent and renowned trilobite species.

IN MEMORY OF . . .

I would like to take this opportunity to fondly remember Terry Abbott, Tom Whiteley, George Lee, Riccardo Levi-Setti, Eugene Thomas, Richard Heimlich, Fred Wessman, Lloyd Gunther, and Patrick Bommel, each of whom served to ignite, inspire, and nurture my interest in trilobites. These exceptional teachers, collectors, and explorers have all sadly passed away in the years since initial work on this book began. My life and the trilobite world in general are both greatly lessened by their absence.

Glossary

Amid the flurry of essays, reports, descriptions, and definitions that fill this trilobite-themed tome, a few words will certainly be less than instantly familiar to anyone who is not a stalwart invertebrate aficionado. This abbreviated glossary may prove beneficial to those of you struggling with an occasional tongue-twisting trilo-term.

appendages: The antennae, limbs, and walking legs that all trilobites possessed that on rare occasions appear as fossilized remnants.

arthropod: The phylum that includes trilobites. All arthropods possess bilateral symmetry, segmented bodies, and chitinous exoskeletons.

axis: The central lobe of a trilobite's three-lobed body design.

benthic: Relating to organisms (including many species of trilobites) that lived at the bottom of a body of water.

Cambrian: The 56-million-year-long period during which trilobites first emerged and attained their greatest global diversity.

EOHARPES BENIGNENSIS (BARRANDE, 1872)
Middle Ordovician, Llandeilian; Dobrotivá Formation; Svatá Dobrotivá, Czech Republic; 2.5 cm

Here is a rarely-seen complete example of this distinctive species, first identified by Barrande in the latter half of the nineteenth century. It has been fossilized with some shell parts still preserved in its negative counterpart.

Carboniferous: Often divided into the Mississippian and Pennsylvanian in North American paleontological parlance, this 60-million-year-long period saw a sharp drop in trilobite speciation.

cephalon: The head of a trilobite, housing the eyes, among other important morphological features.

class: The taxonomic group Trilobita represents the class within which all trilobite orders, families, genera, and species are placed.

compound eyes: The multilensed eyes that often are the most pronounced feature of a trilobite cephalon.

cruziana: Trace fossils comprised primarily of trackways that indicate the walking or crawling lifestyle of a trilobite.

Devonian: The 60-million-year-long period during which trilobite eyes reached their apex of evolutionary design.

dorsal: The top side (back) of a trilobite.

doublure: A cephalic or pygidial rim that stretches under the ventral side of the trilobite.

enrollment: A feature that allowed most trilobite species to assume a ball-like shape to best protect their vulnerable ventral underside.

exoskeleton: The trilobite's hard, calcium-covered outer shell.

fauna: The animals that inhabit a certain region or geological period.

flora: The plant life that inhabits a certain region or geological period.

free cheeks: The parts of a trilobite's cephalon that surround the glabella and can be jettisoned during molting.

genal spine: A spine emanating from the cheek area of the cephalon.

genus: The taxonomic group that ranks above species and below family. *Dalmanites limulurus,* for example, presents the trilobite's genus, followed by its species.

glabella: The midsection of the trilobite cephalon—often convex in shape.

holochroal: A type of trilobite compound eye in which all the lenses are in direct contact and are covered by a single calcite surface.

hydrodynamic: The streamlined shape of some trilobite carapaces that apparently aided them while moving through primeval seas.

hypostome: The trilobite's mouth plate, located on the ventral side of the cephalon.

isopygous: When the trilobite's head and tail are roughly the same size.

Lagerstatte: A location where the fossil content and preservation is of unique scientific importance.

Mesozoic: The Age of Dinosaurs, the 186-million-year-long era that directly followed the Paleozoic.

molting: The action through which a trilobite shed its hard outer shell as an essential part of the growth process.

negative: The reverse side of a trilobite fossil that shows a cast of the positive side of the specimen but retains none (or a minimal amount) of the original calcite shell material.

nonbiomineralized: Lacking the calcite-covering that first emerged with trilobites in the Lower Cambrian; the soft-bodied arthropods found in the Burgess Shale are examples.

opisthothorax: A long, wormlike extension of the thorax that appears on certain Lower Cambrian trilobites and may provide evidence of the trilobites' more primitive antecedents.

Ordovician: The 41-million-year-long period during which some of the largest and strangest trilobite species emerged.

Paleozoic: The 290-million-year-long era that stretched from the Lower Cambrian through the Permian and neatly bookended the rise and eventual fall of trilobites. Paleozoic translates to "the time of ancient life."

pelagic: Relating to organisms (including some species of trilobites) that lived amid the upper layers of a body of water.

Permian: At the end of this 47-million-year-long period, trilobites met their demise, along with 90 percent of the life-forms on Earth.

pleural lobes: The two lobes that flank the central axial lobe; the three together creating a tri-lobe-ite.

prosopon: The finely detailed terracing or structures that can appear on the trilobites' outer shell.

pygidium: The tail end of a trilobite.

schizochroal: A compound eye featuring separate and distinct lenses.

Silurian: The 25-million-year-long period during which predators such as jawed fish began to further erode the trilobites' marine dominance.

species: A group of similar organisms that share common attributes. *Olenellus clarki* presents the trilobite's genus, followed by its species.

spinose: Prominent spines that appear on the trilobite exoskeleton.

suture: Natural breaks in the trilobite carapace that facilitated the molting process.

telson: The pronounced tail spine emanating from the pygidium of certain trilobite species.

thorax: The middle part of the trilobite's anatomy, often featuring between 8 and 15 flexible segments.

ventral: The underside of a trilobite where fossilized evidence of muscle attachment scars (and occasional soft body parts) may be seen.

Index

Page numbers in *italics* represent photos.